Chasing Multinational Interoperability

Benefits, Objectives, and Strategies

CHRISTOPHER G. PERNIN, ANGELA O'MAHONY,
GENE GERMANOVICH, MATTHEW LANE

Prepared for the United States Army
Approved for public release; distribution unlimited

For more information on this publication, visit **www.rand.org/t/RR3068**

Library of Congress Cataloging-in-Publication Data is available for this publication.
ISBN: 978-1-9774-0351-3

Published by the RAND Corporation, Santa Monica, Calif.

Cover: The U.S. Army/Richard Bumgardner,
U.S. Army Europe Public Affairs/flickr.

Preface

This report draws from research and analysis conducted as part of the project *The Value of Interoperability to the Army*, sponsored by the Office of the Deputy Chief of Staff, G-3/5/7, U.S. Army. The purpose of the project was to develop and apply a framework for assessing the benefits of interoperability for the U.S. Army and provide guidance for how best to communicate those benefits to relevant decisionmakers.

This research was conducted within the RAND Arroyo Center's Strategy, Doctrine, and Resources Program. RAND Arroyo Center, part of the RAND Corporation, is a federally funded research and development center (FFRDC) sponsored by the United States Army.

RAND operates under a "Federal-Wide Assurance" (FWA00003425) and complies with the *Code of Federal Regulations for the Protection of Human Subjects Under United States Law* (45 CFR 46), also known as "the Common Rule," as well as with the implementation guidance set forth in DoD Instruction 3216.02. As applicable, this compliance includes reviews and approvals by RAND's Institutional Review Board (the Human Subjects Protection Committee) and by the U.S. Army. The views of sources utilized in this study are solely their own and do not represent the official policy or position of DoD or the U.S. Government.

Contents

Figures and Tables

Figures

Tables

Summary

Recent U.S. national defense policies have focused on the importance of multinational interoperability to meeting U.S. defense goals. However, even with the attention given to interoperability with foreign militaries, the U.S. Army is still not interoperable with whom it wants, when it wants.

There are several reasons that achieving interoperability is an ongoing challenge. We argue that one reason is that policymakers do not have a precise enough understanding of why more and better interoperability is needed. In many ways, "interoperability" is a buzzword often touted as the solution to an unexplained problem. Or worse, as a tautological argument: The need to be interoperable hinges on the fact that, historically, military forces have been rather terrible at being interoperable.

In this report, we use the literature, detailed discussions with military operators and leadership, and several case studies to define the values underpinning interoperability.

Benefits and Objectives

We identify and describe several nonexclusive benefits most often ascribed to interoperability, which are illustrated in Figure S.1. Based on these benefits, we identified the following three overarching objectives for pursuing interoperability, which logically combine the various benefits (the apices of the triangle in Figure S.1):

Figure S.1
Interoperability Benefits, Objectives, and Strategies

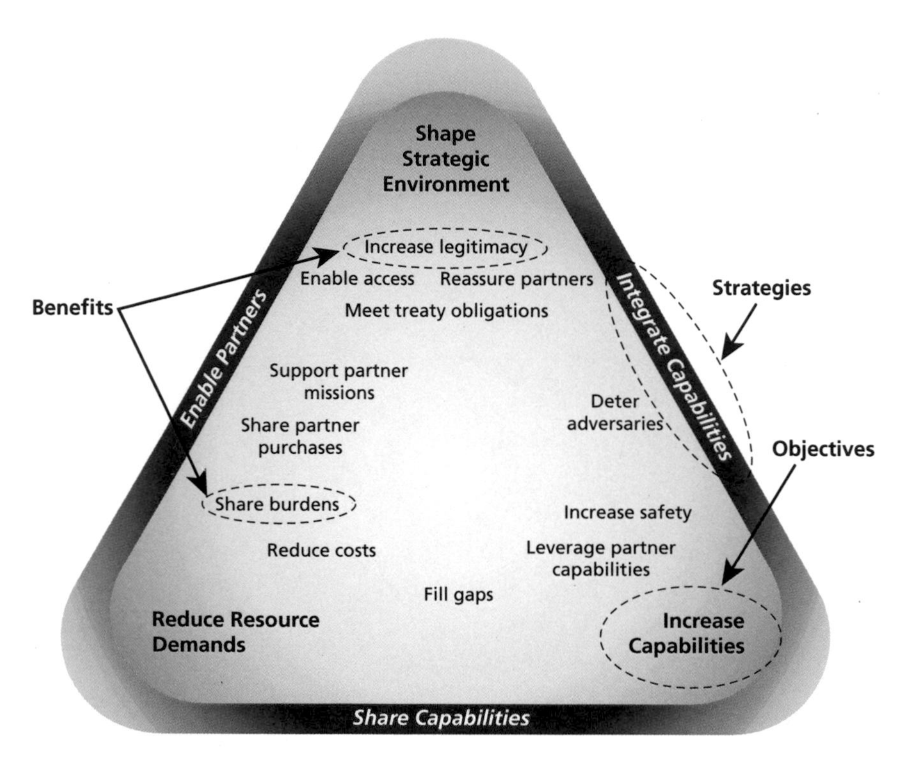

- **Shape the strategic environment**: The United States works with partners and allies to *demonstrate commitment* and *enhance legitimacy*.[1] Interoperability goes beyond simply signing on to a coalition. Operating together on the ground can provide a much more credible signal of intent, reassure partners of continued support,[2] and further build legitimacy in operations.

[1] Nora Bensahel, "International Alliances and Military Effectiveness: Fighting Alongside Allies and Partners," in Risa Brooks and Elizabeth Stanley, eds., *Creating Military Power: The Sources of Military Effectiveness*, Stanford, Calif.: Stanford University Press, 2007.

[2] This belief underpins almost all alliance structures and is at the core of the North Atlantic Treaty Organization (NATO)'s principle of collective defense and efforts to enhance interoperability. NATO commitment increases partner reliability, but such commitments do not fully remove the possibility that a partner might renege on any agreement. Brett Ashley

- **Increase multinational capabilities**: The United States and its partners and allies pool resources to access *greater operational capacity* and *more-effective combined capabilities*.[3] Capabilities can be massed at greater scale by building the necessary interoperability among forces. Partners can also provide capabilities that are superior to what the U.S. Army can provide on its own or are available on faster timelines, therefore filling important gaps in U.S. force structure and posture. By operating directly and effectively with partners, rather than simply deconflicting operations with partners, the U.S. Army can increase safety, such as by reducing fratricide and collateral damage.
- **Reduce resource demands**: The United States can reduce costs by increasing *burden-sharing* with allies and partners. This can take many forms. Joint acquisition programs can distribute research and development expenses. Partners can share support services, taking advantage of scale and reducing unnecessary duplication. And finally, partners can undertake activities in place of the United States.

The three broad interoperability objectives are often combined in subtle ways to rationalize investments in interoperability. We describe several current cases where the United States and its partners are building interoperability. We found that, for most cases, those developing or promoting these capabilities can argue that they fulfil at least two of the three objectives, constituting three interoperability investment strategies. These strategies—integrate capabilities, share capabilities, and enable partners—correspond to the sides of the interoperability values triangle in Figure S.1.

Leeds, "Alliance Reliability in Times of War: Explaining State Decisions to Violate Treaties," *International Organization*, Vol. 57, No. 3, Autumn 2003. Investing in interoperability can increase partners' reliability.

[3] George Liska, *Nations in Alliance*, Baltimore, Md.: Johns Hopkins University Press, 1962, p. 26.

Weighing the Risks and Resources

Interoperability comes with risks and entails resources above and beyond what a force would do for single-country operations. Risks can include disagreements among partners over strategic objectives or how to achieve those objectives; complex and at times overbearing national caveats that forces might claim; compatibility issues, whether they are human, procedural, or technical, that are baked into operations and impossible to overcome on the fly; and the common and cumbersome command and control processes that come with multinational operations to maintain sovereignty and allocate decisionmaking power.

The unclear and often unstated benefits of interoperability are similarly shown in how it is resourced. Bottom-up efforts to bring forces together in the hopes of building interoperability are levied on already existing processes and activities, often with additional costs and in competition with other activities. Therefore, tactical units bear the brunt of interoperability demands. In this case, interoperability is used as a justification without clear value. In the future, it will be necessary to have a clearer understanding of the benefits interoperability brings, as well as mechanisms for properly inserting those requirements within units and commands.

Findings and Recommendations

The benefits ascribed to interoperability vary widely and are not always well articulated or matched to the types of interoperability that are developed. The United States works closely with foreign militaries to close capability gaps, shape the strategic environment, and reduce resource demands when meeting national interests. Potential benefits from interoperability are not always achievable and are highly context-dependent. The choices of partners, scenarios, and functions with and in which to be interoperable are dynamic.

From this look at the benefits that interoperability might bring and how such benefits are constructed in a cross-section of the Army's interoperability initiatives, we offer the following potential next steps for the Army:

- Do not assume high levels of interoperability are valuable, or even possible, with most partners. One size does not fit all.
- Define requirements for interoperability based on the benefits that interoperability can provide.
- Be specific about the benefit expected from any given investment to build interoperability, and develop metrics to evaluate interoperability outcomes.
- Do not include interoperability as an objective in strategic documents (including country plans), unless the purpose of interoperability is clearly defined in the document in terms of end state and benefits.
- Examine U.S. Army processes for obstacles to building interoperability.
- Assign a proponent for interoperability at the right level.
- Commit to interoperability as part of an enduring partnership with and collaboration among partners.

Understanding the benefits is one early and important step in moving the U.S. Army forward on interoperability. Building multinational interoperability brings costs and risks that will also need to be weighed as the Army competes against other capabilities in a resource-constrained environment. Aiming the institutions in the right direction and understanding why interoperability needs to be developed will be instrumental in eventually defining the benefits thereof and building the force of the future.

Acknowledgments

We are grateful to HQDA G-3/5/7 for sponsoring this study; we particularly appreciate the support of MG William Hix and MG Christopher McPadden. We thank COL Joseph Fossey, COL Robert Howieson, and Mark McDonough, HQDA G-3/5/7 DAMO SSC, for monitoring the study and for providing frequent and constructive feedback during it.

We are grateful to the many people throughout the Department of Defense who were willing to meet with us throughout this study. In particular, we would like to thank the following people who made these meetings possible: COL Matthew Morton and Gus Blum (U.S. Army Central); Leonardo Hernandez and Justino Lopez (U.S. Army North); COL Howard Kirk and Susan Groth (U.S. Army Africa); COL James Dodson and BG Edmundo Villarroel (U.S. Army South); LTC David Gordon and Michael Tedesco (U.S. Army Europe); Don Birdseye, COL Peter Don, Rodney Laszlo, CW5 Roy L. Rucker, Sr., LTC Josh Stephenson, and Christopher Zaklan (U.S. Army Pacific); BG Ollie Kingsbury (82nd Airborne Division); LTC Joel Gleason (7th Army Training Command); Brigadier James J. Learmont (U.S. Central Command); James Mitre (Office of the Secretary of Defense for Policy); and LTC Christopher Cline (Headquarters, Department of the Army, G-3/5/7).

At RAND, we thank Timothy Bonds and Sally Sleeper for their support throughout the study. In addition, we thank Catherine Dale, Michael Gaines, Jakub Hlavka, Jennifer Moroney, Maritta Tapanainen, Steven Popper, and Paul Steinberg for their help. Emily-Kate Chiusano and Holly Johnson provided assistance throughout the study. We also would like to thank those who provided access to data or comments on the research: Katharina Best, Gian Gentile, John Gordon, Beth Grill, Bryan Hallmark, Benjamin Harris, Mike Linick, Michael McNerney, Shanthi Nataraj, and J. D. Williams.

We thank our two peer reviewers, Chad Serena (from RAND) and Caitlin Talmadge (from Georgetown University), who provided ample critical feedback.

Abbreviations

ABCANZ	American, British, Canadian, Australian, and New Zealand
ASCC	Army service component command
COP	common operational picture
DoD	U.S. Department of Defense
DLD	Digital Liaison Detachment
GCC	geographic combatant command
HQDA	Headquarters, Department of the Army
MN	multinational
NATO	North Atlantic Treaty Organization
NDS	National Defense Strategy
ROK	Republic of Korea
SC	security cooperation
UN	United Nations
USARAF	U.S. Army Africa
USARCENT	U.S. Army Central
USAREUR	U.S. Army Europe

USARNORTH	U.S. Army North
USARPAC	U.S. Army Pacific
USARSOUTH	U.S. Army South
VJTF	Very High-Readiness Joint Task Force

CHAPTER ONE

Introduction

Background

In modern discussion about warfare and military capabilities, interoperability between organizations almost invariably comes up. This might include interoperability within a service (e.g., Army aviation working alongside Army fires), among services (e.g., Army and Air Force units working together) and other governmental agencies (e.g., the Army working with the State Department), or internationally (e.g., the Army working with allies and partners). There is a significant body of literature on these types of interoperability, with the common refrain that *interoperability is critical, and more and better interoperability is needed.*[1]

This refrain is reflected in the U.S. Department of Defense (DoD)'s recent focus on interoperability. The 2014 National Military Strategy and prior Quadrennial Defense Reviews all mentioned interoperability, often associated with similarly advanced states and the North Atlantic Treaty Organization (NATO). The 2018 National Defense Strategy (NDS)[2] picks up that interest, highlighting the importance of alliances and partnerships to meeting U.S. defense goals. Enhancing

[1] Patricia A. Weitsman, *Waging War: Alliances, Coalitions, and Institutions of Interstate Violence,* Stanford, Calif.: Stanford University Press, 2014; Olivier Schmitt, *Allies That Count: Junior Partners in Coalition Warfare,* Washington, D.C.: Georgetown University Press, 2018.

[2] DoD, *Summary of the 2018 National Defense Strategy of the United States of America: Sharpening the American Military's Competitive Edge*, Washington, D.C., January 1, 2018.

ties with allies and partners represents one of three NDS lines of effort, along with making the force more lethal and reforming DoD.

Over the years, military leadership's push for more multinational interoperability has led to military activities full of partner involvement. Large-scale combined training, both by policy and practice, entails working with multinational partners.[3] The U.S. military education system reserves slots for members of the Foreign Service, providing them a glimpse of U.S. military theory and practice.[4] There are also programs to insert Foreign Service members, either as liaisons or as members of staff sections, into U.S. military units.[5] These are just a few of many activities under the rubric of "security cooperation" (SC) that the joint force designs to foster unit-to-unit relationships, cultural understanding, compatibility in equipment, and other components of interoperability.[6] This interest has grown of late. As recently as 2013, the guiding Army regulation on SC did not mention interoperability directly,[7] but the 2015 handbook version of that regulation, which includes descriptions of many SC-related programs, mentions it more than 100 times.[8]

---

[3] Center for Army Lessons Learned, "Commander's Guide to Multinational Interoperability," Fort Leavenworth, Kan.: U.S. Army Combined Arms Center, No. 15-17, 2015.

[4] See Angela O'Mahony, Thomas S. Szayna, Christopher G. Pernin, Laurinda L. Rohn, Derek Eaton, Elizabeth Bodine-Baron, Joshua Mendelsohn, Osonde A. Osoba, Sherry Oehler, Katharina Ley Best, and Leila Bighash, *The Global Landpower Network: Recommendations for Strengthening Army Engagement*, Santa Monica, Calif.: RAND Corporation, RR-1813-A, 2017; Christopher G. Pernin, Angela O'Mahony, Thomas S. Szayna, Derek Eaton, Katharina Ley Best, Elizabeth Bodine-Baron, Joshua Mendelsohn, and Osonde A. Osoba, "What Is the Global Landpower Network and What Value Might It Provide?" *Defense and Security Analysis*, Vol. 33, No. 3, 2017.

[5] O'Mahony et al., 2017; Pernin et al., 2017. See, for instance, the Foreign Liaison Officer Program, Military Personnel Exchange Program, and Cooperative Personnel Program.

[6] Jennifer D. P. Moroney, David E. Thaler, and Joe Hogler, *Review of Security Cooperation Mechanisms Combatant Commands Utilize to Build Partner Capacity*, Santa Monica, Calif.: RAND Corporation, RR-413-OSD, 2013.

[7] Army Regulation 11-31, *Army Security Cooperation Policy*, Washington, D.C.: Headquarters, Department of the Army, March 21, 2013.

[8] The 2015 handbook version was significantly (roughly 100 pages) longer than the 2013 regulation, owing to a compendium of specific programs listed in the back. See Army

However, even with the renewed and growing attention and decades of lingering interest in interoperability, only a fraction of these efforts to create interoperability truly build long-term solutions and achieve the type of unit-to-unit trust and compatibility essential for multinational operations.[9] This result is clearly seen in the Army, which is not interoperable with whom it wants, when it wants.

There are various reasons why achieving interoperability is an ongoing challenge. First, *it is often not clear, or at least not easily described, how much interoperability is needed, with which partners, and for what reasons.* Lessons from recent operations highlight areas where greater interoperability would be beneficial.[10] However, Army planning and tactical unit training tend not to incorporate partner capabilities—executing missions with a partner is rarely a consideration for unit readiness.[11] As a result, how the Army can leverage interoperability to meet its objectives is not integrated into how the Army prepares to accomplish those objectives.

Second, *it is unclear what steps are necessary for two countries to be interoperable.* As mentioned, the U.S. military has ample activities that bring U.S. forces closer together with foreign forces. However, interoperability is not the only objective of these activities, and how best to engage with partner units is often left to the tactical units that then need to fit interoperability in alongside their many other demands. Targeting partner engagement activities to more effectively build and sustain interoperability will require more focused assessment, monitor-

Pamphlet 11-31, *Army Security Cooperation Handbook*, Washington, D.C.: Headquarters, Department of the Army, February 6, 2015.

[9] Christopher G. Pernin, Jakub P. Hlavka, Matthew E. Boyer, John Gordon IV, Michael Lerario, Jan Osburg, Michael Shurkin, and Daniel C. Gibson, *Targeted Interoperability: A New Imperative for Multinational Operations*, Santa Monica, Calif.: RAND Corporation, RR-2075-A, 2019.

[10] Center for Army Lessons Learned, 2015.

[11] Observation based on our interviews with Army service component command (ASCC) and unit personnel.

ing, and evaluation of Army activities, as well as better guidance and support for tactical units.[12]

Third, *interoperability is an investment that competes for the interest of leadership and financial resources with other priorities and capabilities.* Interoperability is rarely a foreign policy goal; instead, it is often relegated to a by-product of other investments and activities. When interoperability is the focus of an investment proposal, it competes with other resource needs for people, equipment, and training. The more that is known about what value a capability offers, the easier it is to make the argument for investing.

Finally, *the benefits of interoperability relative to its costs and risks are often not well understood.* Not knowing the value of interoperability limits the funding and senior leader interest required to build it. It is not clear whether the benefits of increased interoperability outweigh its costs, primarily in the form of increased strategic or operational dependence on partner forces, expenses related to making training and exercises multinational, requirements for compatible equipment, and political friction when disagreements emerge in peacetime or conflict. When the benefits from interoperability are not easily understood or conveyed, interest in expending time or resources will be limited. As a first step, understanding the benefits will help determine just how much and with whom interoperability should be built.

Taken together, these challenges to interoperability reflect the fact that policymakers do not have a precise enough understanding of why *more and better interoperability is needed.* In many ways, "interoperability" is a buzzword often asserted as the solution to an unexplained problem. Or worse, as a tautological argument: the need to be interoperable hinges on the fact that, historically, military forces have been rather terrible at doing so. Understanding the benefits will help constitute a compelling vision for improving the Army's ability to meet the needs of the nation. Deeper understanding of how the benefits

---

[12] Angela O'Mahony, Ilana Blum, Gabriela Armenta, Nicholas Burger, Joshua Mendelsohn, Michael J. McNerney, Steven W. Popper, Jefferson P. Marquis, and Thomas S. Szayna, *Assessing, Monitoring, and Evaluating Army Security Cooperation: A Framework for Implementation,* Santa Monica, Calif.: RAND Corporation, RR-2165-A, 2018.

fare against the costs and risks and against competing investments can ensue.

Objectives

We developed a framework for this study for more precisely assessing the benefits often ascribed to interoperability. We illustrate how that framework could work by providing examples of recent Army initiatives and discussing some of the resource issues involved in generating and using interoperability.

We drew on the interoperability literature, in an attempt to capture some of the sentiment and assumptions being made about the value of interoperability. The literature is cited throughout. We drew predominantly on stakeholder interviews to formulate the current perceptions of and expectations for interoperability. These interviews generally followed the question protocol shown in the appendix. We targeted individuals in the Pentagon, in the combatant commands and service component commands, and other individuals in outlying areas that had an expected role in implementing interoperability. We focused on both the owners of programs and planning staff who use and make judgments about the value of interoperability. Because of the occasionally nebulous nature of the term *interoperability*, it is not easy to identify the most knowledgeable individual in a given office; thus, the team relied on snowball interviews to help capture subjects. All told, we discussed interoperability with nearly four dozen individuals. Because of the nature of the subject, not all individuals are cited in this document, nor are names or affiliations shared more broadly.

As a precursor to determining the benefits of interoperability, we first need to define what we mean by interoperability. The military definition of interoperability has varied over time. At times, the definition has focused on technical aspects of systems or platforms operating together, such as through the exchange of data over communication links.[13] Other times, the definition has taken on a broader, more operational, and strategic flavor, focusing on enabling units or nations to operate in "synergy"[14] with partners. An older definition, originally used in versions of Joint Publication 1-02,[15] has gained prominence because of its utility across technical, operational, and strategic issues:

> The ability of systems, units, or forces to provide services to and accept services from other systems, units, or forces and to use the services so exchanged to enable them to operate effectively together.

This definition can be understood simply as the ability of units to consolidate capabilities through common or similar technical systems and nontechnical processes. If an Army ground unit needs fire support during an operation, for instance, achieving interoperability (using this definition) means that the unit can call for fires from partner nation forces and that those forces can effectively provide fire support in a timely manner. The definition may be applied to any warfighting function: providing logistical support, sharing information, or being able to task-organize with maneuver units from different nations for an exchange of services. Precisely how well interoperability works depends on the human, procedural, and technical connectivity needed and achieved in each case.

Our previous work highlighted the importance of senior leadership better defining *with whom* and *for what* interoperability should be built as a way to better align resources and leadership attention.[16]

---

[13] Thomas Ford, John Colombi, Scott R. Graham, and David R. Jacques, "A Survey on Interoperability Measurement," *Proceedings of the 12th International Command and Control Research and Technology Symposium*, Newport, R.I., June 19–21, 2007.

[14] Joint Publication 1-02, *DoD Dictionary of Military and Associated Terms*, Joint Chiefs of Staff: Washington, D.C, 2010.

[15] This definition of *interoperability* is no longer part of Joint Publication 1-02 but can be found in its previous versions, including the one from 1994 through January 2000—see Joint Publication 1-02, *DoD Dictionary of Military and Associated Terms*, Joint Chiefs of Staff: Washington, D.C., January 10, 2000.

[16] Pernin et al., 2019.

A significant missing part, we found, was that, aside from a couple of examples, few units are being given the specific guidance necessary to build interoperability. One example of a top-down effort is the partnership between the U.S. 82nd Airborne Division and the British 16th Air Assault Brigade, which are aiming to build a capability to conduct high-end airborne operations with what is intended to be "seamless" integration.[17] In many ways, this targeted effort was a departure from the past interoperability aspired to by airborne units and brought tactical units closer than ever before. While targeted interoperability of this type is rare, interest in it seems to be growing as more nations interact at lower echelons on shorter timelines than in the past.[18]

Organization of This Report

We focus on benefits in this report as a first step toward understanding the overall value proposition of interoperability. In Chapter Two, we describe various benefits often ascribed to interoperability. In Chapter Three, we illustrate those potential benefits with examples. Understanding the overall value proposition—both benefits *and* costs—will inevitably need to include a much better articulation of the costs and risks of interoperability, both of which are not well described in the literature or captured widely in practice. Nonetheless, we provide some observations and framing of the risks and resource demands of interoperability in Chapter Four as a starting point for further development. The concluding chapter offers perspectives about the Army's approach to interoperability going forward. An appendix includes questions used during our interviews.

---

[17] Christopher G. Pernin, Katharina Ley Best, Matthew E. Boyer, Jeremy M. Eckhause, John Gordon IV, Dan Madden, Katherine Pfrommer, Anthony D. Rosello, Michael Schwille, Michael Shurkin, and Jonathan P. Wong, *Enabling the Global Response Force: Access Strategies for the 82nd Airborne Division*, Santa Monica, Calif.: RAND Corporation, RR-1161-A, 2016.

[18] Pernin et al., 2019.

CHAPTER TWO

The Possible Benefits of Interoperability

In this chapter, we develop a framework to help policymakers understand more precisely the possible benefits of investments in interoperability. We first identify objectives that interoperability can help accomplish and then discuss strategies to match investments in interoperability to these objectives.

Interoperability Benefits, Objectives, and Strategies

Interoperability is best understood as a *means to some other end*, not as an end in and of itself. Interoperability, therefore, is beneficial for what it allows multinational military forces to accomplish. To identify discrete benefits that might accrue through interoperability, we conducted interviews, reviewed the literature on interoperability, and examined leadership statements. This included reading national security strategy documents, bilateral and multilateral strategic vision statements, and Army operational guidance; reviewing Army lessons learned from past and ongoing operations; and conducting interviews at ASCCs; some geographic combatant commands (GCCs); Headquarters, Department of the Army (HQDA); and operational units.[1] All told, we collected a list of the following benefits often ascribed to interoperability:

[1] We describe our approach in more detail in the appendix.

- *Enabling access to locations and populations:* There is uncertainty in where U.S. forces might operate for future operations. Interoperability can make it easier to work out operational details of access.
- *Leveraging partner capabilities:* Some partners have valuable niche capabilities that can bolster overall U.S. Army performance.
- *Filling capability gaps in force structure:* The U.S. Army has force structure and capability gaps in key scenarios that partners could help bridge.
- *Increasing legitimacy of operations:* The U.S. Army often seeks involvement from partners to show commitment and enhance legitimacy of its operations.
- *Increasing operational safety (decrease fratricide, collateral damage):* The United States will inevitably work together with partners and thus needs to reduce downside effects of operating with disparate forces, such as fratricide and collateral damage.
- *Deterring adversaries:* By increasing capabilities and demonstrating commitment, interoperability can deter adversaries.
- *Meeting treaty obligations:* Interoperability increases multinational capabilities to meet treaty obligations.
- *Reassuring partners:* Working closely with partners helps partners to understand U.S. Army capabilities and demonstrates U.S. commitment.
- *Reducing costs of operations:* Global commitments over long periods entail finding ways of reducing overall costs of operations. Interoperability can help efforts to maintain readiness for future operations while meeting current demands.
- *Shaping partner purchases:* Interoperability increases purchases of shared materiel and training.
- *Sharing burdens for operations:* Interoperability provides a mechanism for burden-sharing.
- *Supporting partner-led missions:* The United States is committed to supporting partners in maintaining stability and sovereignty.

This list is not exhaustive, nor are the benefits listed here mutually exclusive—for example, efforts to deter adversaries are also likely to reassure partners and to help meet treaty obligations. However, the list is representative of the interoperability benefits that appeared most frequently in our research. The breadth and diversity of outcomes that interoperability might facilitate makes the benefits of interoperability contextually specific to each situation. Moreover, the perceived benefits of interoperability reflect the concerns of each observer. For example, not surprisingly, operators we interviewed were more likely to flag the value of interoperability in reducing fratricide, while strategists highlighted the importance of deterring adversaries.

It is important to note that while we provide a synthesis of the benefits ascribed to interoperability, we do not assess the extent to which interoperability accomplishes these benefits. Some of these benefits defy obvious or well-accepted assignment of value. As an example, deterrence is widely studied but not easily quantified in terms of how much is necessary for what expected results. Therefore, assessing the beneficial impact of interoperability on deterrence is similarly fraught. Other benefits, while theoretically possible, might not be easily identified in practice. For example, if the aimed benefit of interoperability is to reduce U.S. military force structure and replace it with a partner capability, that reduction would need to be found as an avoided cost in the complex planning and programming that determines the force. The articulation of the possible benefits of interoperability should be seen as the beginning, rather than the end, of a conversation that assesses the full costs and risks. Once the possible objectives for interoperability are identified, criteria for conducting a cost-benefit analysis follows.

Based on this list, we identified three overarching objectives for pursuing interoperability: *shape the strategic environment*, *increase multinational capabilities*, and *reduce resource demands*.

The triangle in Figure 2.1 maps the exemplar interoperability benefits noted previously, in terms of these three overarching interoperability objectives. Notably, many of the benefits we identified encompass multiple objectives, which we place between the three overarching objectives. For example, interoperability helps deter adversaries by both *shaping the strategic environment* and *increasing the overall capabilities* of partnered forces, so that benefit is positioned between those two objectives in the triangle.

Figure 2.1
Mapping Benefits to Interoperability Objectives

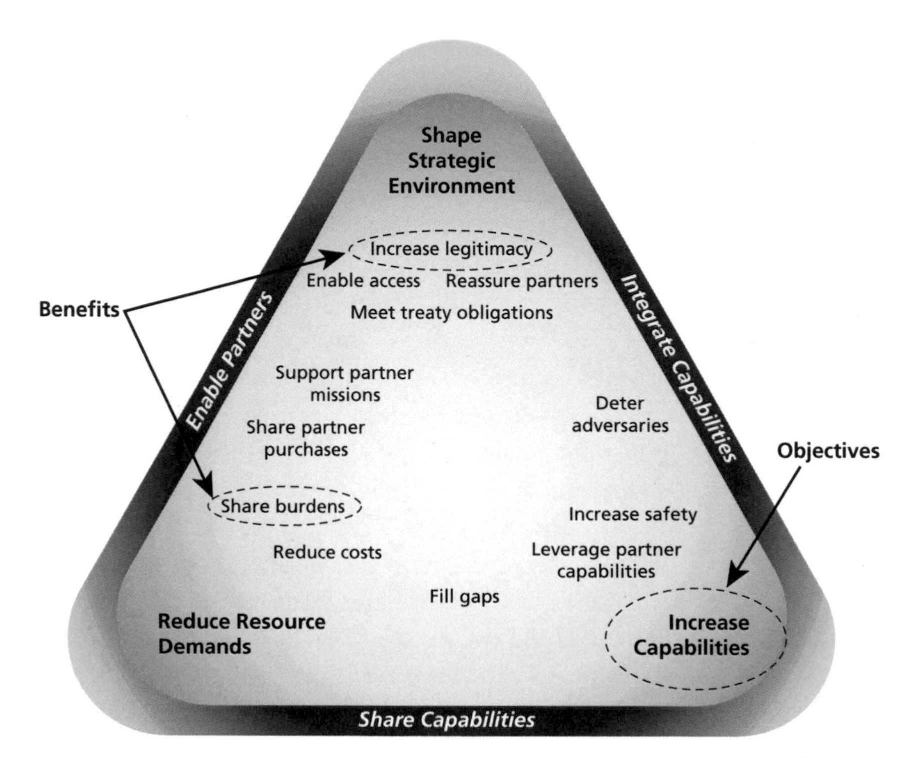

Interoperability Can Shape the Strategic Environment

The United States works with partners and allies to *demonstrate commitment* and *enhance legitimacy*.[2] Interoperability goes beyond simply signing on to a coalition. Operating together on the ground—having "skin in the game"—can provide a much more credible signal of intent, reassuring partners of continued support.[3] Interoperability can

[2] Nora Bensahel, "International Alliances and Military Effectiveness: Fighting Alongside Allies and Partners," in Risa Brooks and Elizabeth Stanley, eds., *Creating Military Power: The Sources of Military Effectiveness*, Stanford, Calif.: Stanford University Press, 2007.

[3] This belief underpins almost all alliance structures and is at the core of NATO's principle of collective defense and efforts to enhance interoperability. NATO commitment increases partner reliability, but such commitments do not fully remove the possibility that a partner might renege on the agreement. Brett Ashley Leeds, "Alliance Reliability in Times of War:

also build legitimacy in operations. Gulf War planners understood the political and strategic importance of having the military forces of Gulf partner states, rather than U.S. forces, liberate Kuwait City.[4] International Security Assistance Force planners recognized the importance of working closely with forces from Muslim-majority countries to maintain the Afghanistan operation's legitimacy with both international and Afghan audiences.[5] U.S. activities with the Republic of Korea (ROK) consistently demonstrate the United States' commitment to fulfilling its treaty obligations by supporting and enabling the defensive operations led by the ROK military forces.[6] Our recent interviews emphasized the importance of working with regional partners for gaining acceptance and access. Partner forces can legitimately go places and execute tasks where U.S. forces would be less accepted by local populations.

Interoperability Can Increase Multinational Capabilities

The United States and its partners and allies pool resources to access *greater operational capacity* and *more-effective combined capabilities.*[7] We identified three key benefits that partner capabilities provide for the Army. First, when partner capabilities are similar to the U.S.' own capabilities, the United States and its partners can mass capabilities at greater scale by building the necessary interoperability among forces. Examples here might be system-focused, such as amassing Patriot air-defense capabilities across multiple countries, or they can be focused on broader capabilities, such as building interoperability between the U.S.

Explaining State Decisions to Violate Treaties," *International Organization*, Vol. 57, No. 3, Autumn 2003. Investing in interoperability can increase partners' reliability.

[4] Michael R. Gordon, and Bernard E. Trainor, *The Generals' War: The Inside Story of the Conflict in the Gulf*, Boston, Mass.: Little Brown and Co., 1995.

[5] Katharina P. Coleman, "The Legitimacy Audience Shapes the Coalition: Lessons from Afghanistan, 2001," *Journal of Intervention and Statebuilding,* Vol. 11, No. 3, 2017.

[6] Don Oberdorfer and Robert Carlin, *The Two Koreas: A Contemporary History,* New York, N.Y.: Basic Books, 2014.

[7] George Liska, *Nations in Alliance*, Baltimore, Md.: Johns Hopkins University Press, 1962, p. 26.

82nd Airborne Division and the British 16th Parachute Regiment for global response force missions, which increases U.S. and British capacity to respond rapidly to crises.[8]

Second, partners can provide capabilities that are superior to what the Army can provide on its own or are available on faster timelines, thus filling important gaps in U.S. force structure and posture. For example, NATO bridging units are important enablers to possible future warfighting in Europe, yet such units are in short supply on timelines applicable in that region.[9] Similarly, information collection often favors local or regional forces who might have a much better understanding of the environment, people, and even transportation networks. Some partners may also bring greater capabilities to operate in certain environments, such as mountains, jungles, and the Arctic.

Finally, operating directly and effectively with partners, rather than simply deconflicting operations with partners, can reduce operational safety risks, such as fratricide and collateral damage.

Interoperability Can Reduce Resource Demands

The United States can reduce costs by increasing *burden-sharing* with allies and partners. This can take many forms. Joint acquisition programs can distribute research and development expenses. Partners can share support services—taking advantage of scale and reducing unnecessary duplication. As a command tasked with executing missions in a geographically large area of responsibility and fewer assigned forces than other theaters, personnel at U.S. Army Africa (USARAF) look to pool resources with allies and partners. For example, the United States provides airlift capabilities to France, while France provides medical support to many U.S. exercises.[10] Finally, partners can undertake activities instead of the United States. In South America and Africa, much of the training the U.S. Army provides builds partners' capacity to par-

---

[8] Pernin et al., 2016.

[9] David A. Shlapak, and Michael Johnson, *Reinforcing Deterrence on NATO's Eastern Flank: Wargaming the Defense of the Baltics*, Santa Monica, Calif.: RAND Corporation, RR-1253-A, 2016.

[10] U.S. Army Africa personnel, interview with the authors, Vicenza, Italy, March 2018.

ticipate in multinational peacekeeping operations—tasks that the U.S. Army might undertake if capable partners did not exist.[11]

Matching Interoperability Objectives to Interoperability Investment Strategies

The three broad interoperability objectives mentioned previously—shaping the strategic environment, increasing capabilities, and reducing resourcing demands—are often combined in subtle ways to rationalize investments in interoperability. In our look through current cases, we found that, for most cases, at least two of the three objectives are often argued.[12] In this section, we describe three interoperability investment strategies we observed; each entails blending two of the interoperability objectives. These strategies—integrate capabilities, share capabilities, and enable partners—correspond to the sides of the interoperability benefits triangle in Figure 2.2.

Integrate Capabilities: Interoperability to Accomplish Strategic and Capability Objectives

In 2017, Ben Hodges, former commander of USAREUR, posited that 2018 would be "the year of integration" for NATO allies.[13] He identified three technical areas in which integration was particularly important: tactical radios, digital fires, and a common operating picture. NATO operations in Europe exemplify the nexus between shaping the strategic environment and increasing capabilities, with a focus on large-scale operations against a near-peer adversary. The United States and its partners and allies are deepening interoperability both to signal resolve and to develop needed capabilities. This interoperability strat-

---

[11] Mindy Anderson, "U.S. Army Africa 'Train the Trainers' in Ghana," U.S. Africa Command, July 1, 2011; Nastasia Barcelo, "Uruguay and the U.S. Train to Enhance Peacekeeping Missions," *Diálogo Digital Military Magazine*, August 14, 2015.

[12] In the case of combined U.S.-ROK military capabilities, arguments for all three interoperability objectives were articulated.

[13] David Vergun, "Army, Allies Strive for Greater Interoperability in Europe," *Army News Service*, October 18, 2017.

Figure 2.2
Interoperability Investment Strategies Related to Interoperability Objectives

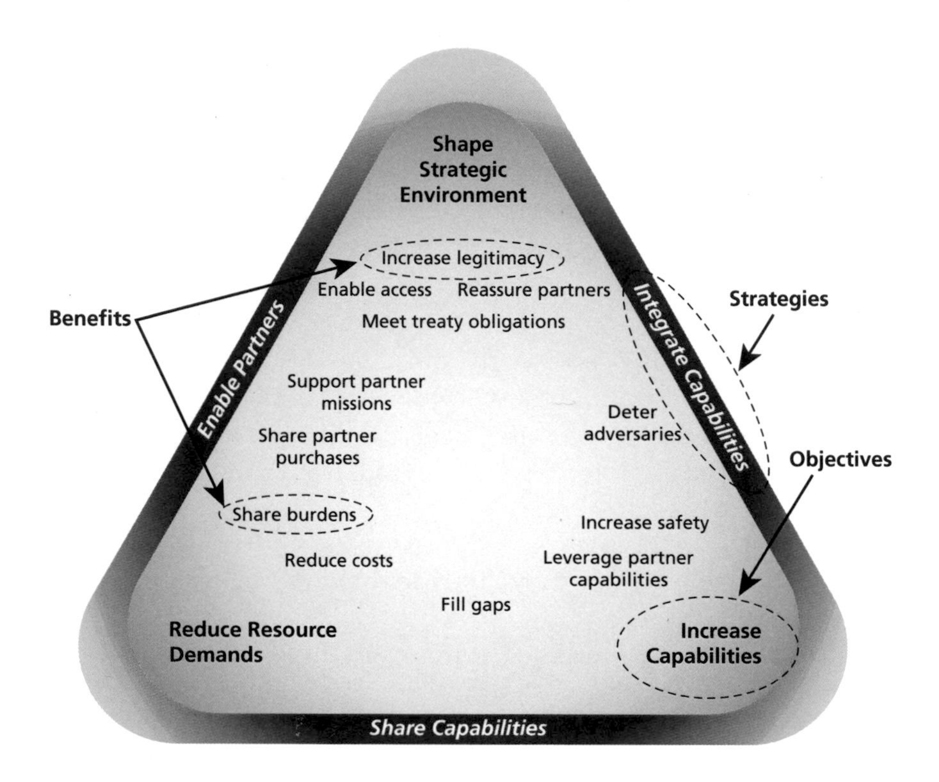

egy is beneficial for contexts in which there is a strategic rationale for working with partners, increased capabilities are operationally necessary, and the operations are high priorities that justify the resources necessary for building and sustaining such joint efforts.

Interoperability investments should focus on mechanisms to enable integration, such as introducing joint training; improving joint command, control, communications, computers, and intelligence capabilities; making materiel interoperable; understanding; and, when appropriate, standardizing procedures.

Share Capabilities: Interoperability to Accomplish Capabilities and Resourcing Objectives

Many Army tasks can also be performed by partners and allies. From an interoperability perspective, the most important task characteristic is that the task gets done to a specified standard—not who does it. There are many examples of such capabilities: strategic lift, demining, perimeter security, medical support, information collection, sustainment, and transportation. Deciding which partner performs this type of activity might depend on which partner can do it more easily or faster: Who has readily available resources, or can do it more effectively? This interoperability strategy is valuable in contexts in which there is not a strong strategic rationale for visibly working with partners. In contrast, capabilities are needed to accomplish the task, and reducing resourcing demands can free up resources for other, more highly prioritized uses.

To support this type of interoperability, the Army should invest in additional capacity in areas where it can provide support to partners. Conversely, the Army can potentially reduce investments in areas where it can expect support from its partners.

Enable Partners: Interoperability to Accomplish Strategic and Resourcing Objectives

Not all Army activities with partners deliberately result in interoperability. One priority for Army security cooperation is to build partners' capacity to meet objectives that both the partner and the United States share—such as strengthening the partner's internal defense or preparing the partner for multinational peacekeeping operations. While the partner and the United States may operate together in the future, such interoperability is a secondary consideration. Similarly, while the training and equipment that the United States provides to partners may be interoperable with that used by the Army, this is more a reflection of best practices and U.S. expertise and processes than a first-order objective of those interactions.

This interoperability strategy is beneficial for contexts in which there is not a strong need for increased capabilities to augment U.S. operations and in which the United States has a strategic rationale for supporting its partner. U.S. resourcing demands could be reduced if partners participate in operations that the United States might otherwise have had to undertake—such as regional stability operations or demining. As laid out in the 2018 NDS, building partner capacity remains a priority for the United States. Interoperability investments for this type of activity fulfill the United States' strategic interest in a globally expanded future force. Army investments should focus on strengthening partners' ability to execute activities that are high priorities for the partner and in keeping with U.S. strategic objectives. If the United States plans to work with a partner in the future, generating basic compatibility with U.S. forces can be a useful goal in building partner capacity; however, such an objective is likely to be secondary.

Summary

Figure 2.3 puts together all the pieces of the triangle that make up the value proposition for interoperability—exemplar benefits, interoperability objectives, and investment strategies.

Benefits can accrue from interoperability and inform why investments in interoperability might be made. Interoperability can be a means for shaping the strategic environment, building new capabilities, or reducing future demands on resources. Each of these areas can be combined to aid in explaining and rationalizing the investments necessary for realizing envisioned interoperability.

Figure 2.3
Benefits, Objectives, and Strategies for the Value Proposition for Interoperability

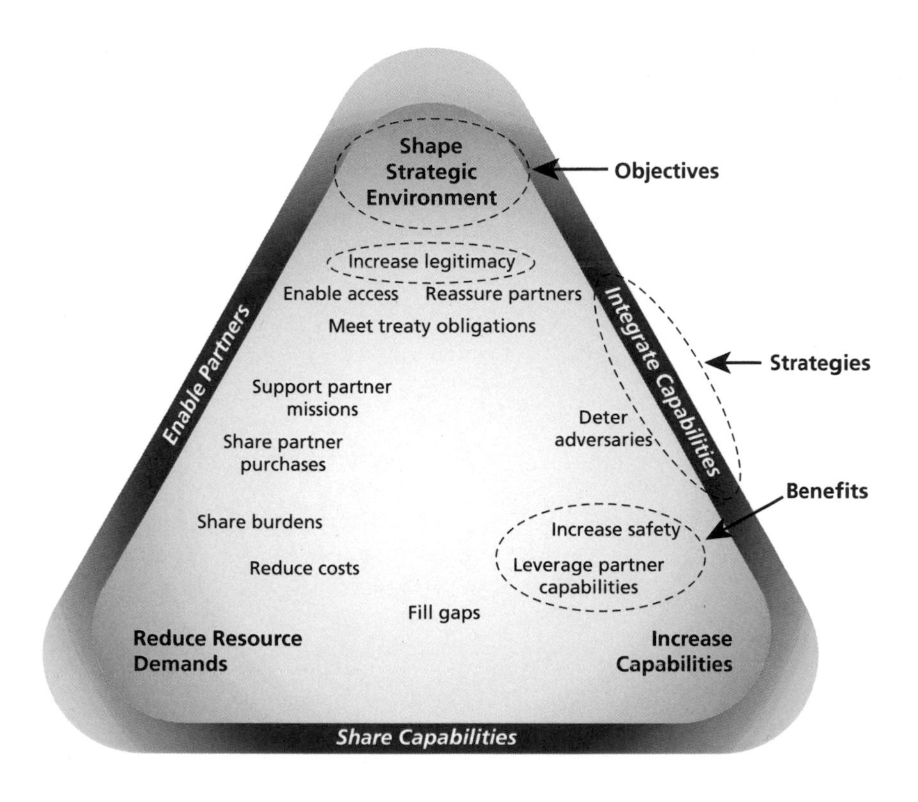

CHAPTER THREE

Interoperability in Practice

Translating the potential benefits of interoperability into a strategy for building it requires understanding how those benefits stem from interoperability that has been built with partners. As we discuss in this chapter, developing an effective interoperability strategy *in practice* requires specifying with whom the U.S. Army will be interoperable, in what functional areas, and for what purposes. This chapter provides insights from ongoing examples of interoperability to highlight how different approaches to building interoperability provide various benefits to U.S. forces.

Designing Interoperability to Match Scenarios, Functions, and Partners

Returning to our definition of interoperability in Chapter One, the "ability to provide services to" a partner force fundamentally involves articulating three parameters: (1) a *scenario* in which interoperability could be realized with those partners and functions, (2) a set of *functions*, and (3) a set of *partners*. Sometimes, discussions about interoperability will start with only one parameter—the desire to build interoperability with a specific partner, or around a specific function, or in a specific scenario—and attempt to be agnostic about the rest. For example, an attempt to build a mission command system that is not specific to the types of operations or partners that might use it could be readily rationalized from past operations, where neither the type of operation nor the full extent of the coalition could be discerned ahead of time. Nonetheless, requirements for interoperability will typically be best supported when those three parameters are readily explained and logically linked to the benefits.

Developing **planning scenarios** that necessitate specific partner capabilities is one way to better describe and analyze *with whom* and *for what* interoperability needs to be built. In some cases, the scenario may pose a strategic need, such as countering a rising Russian influence, which then may have the depth of analysis to support specific partners and functions that are necessary for interoperability. However, not all scenarios are worked out to the level of analytic detail necessary to make investment decisions for accessing specific capabilities from partner nations. The scenario view may also significantly limit the number of partners identified as important. The U.S. military's efforts to reduce risk in planning means that the force-development process preferably looks to U.S. solutions, minimizing the demand for foreign support for planned contingency operations.

The United States can also choose to focus on a **given function**, such as intelligence or mission command. The persistent imperative to access partner nation intelligence and the history of integrating disparate partners into command structures are two examples that may imply the need to create interoperability among these functions. In these cases, the aim is to improve specific, common functions where partners are routinely involved. A functional view may be challenged in cases where partners vary widely in their capabilities, or where the requirements for specific scenarios might waver from region to region. The military's push for a common "mission partner environment" aims to have wide partner inclusion and offer key services necessary to execute mission command functions in a multinational environment. The extent to which this is tested against actual partner capabilities or in specific contexts and scenarios will of course be a critical signpost of its eventual utility.

Finally, interoperability can be borne of a special, existing relationship with **partners**. The United States–United Kingdom relationship routinely involves each country's forces training, exercising, and operating together, often with the subtext of interoperability. However, without the specificity of functions and scenarios, this interoperability may be only temporary or incidental.

It is inevitable that all three views—the partner, the function, and the scenario—are necessary to fully appreciate the benefits that could be accrued, and to the extent that they are known, those investments will be most easily rationalized.

Exploring the Links Among Partners, Functions, and Scenarios—Some Army Examples

To further explore why interoperability is built, we identified several examples across various GCCs in which the U.S. Army is building multinational interoperability. Each of the efforts started in its own way, whether by partner, function, or scenario; additionally, each has a set of values ascribed to it in writing or in practice. While these examples do not capture all the interoperability activities the Army is undertaking, they do reveal the range and character of ongoing Army efforts to build multinational interoperability for responding to specific theater objectives. Figure 3.1 shows the list of examples by the three parameters.

Examples of Partner-Driven Interoperability

Rather than relying on increased interoperability to meet a specific functional or mission-based need, many instances of U.S. interoperability are instead driven by a decision to further improve strategic or operational relationships with specific partners. Such partner-driven interoperability often stems from a special historic relationship, the prospect of working together in the future, or just the happenstance of proximity. Once identified, interoperability can then be built around certain functional or scenario demands, based on the relationship and specific capabilities of the partner. Determining these functional and mission demands therefore highlights how the United States should work with those prioritized partners to maximize the gains from interoperability.

Figure 3.1
Army Examples of Partner-, Function-, and Scenario-Driven Interoperability

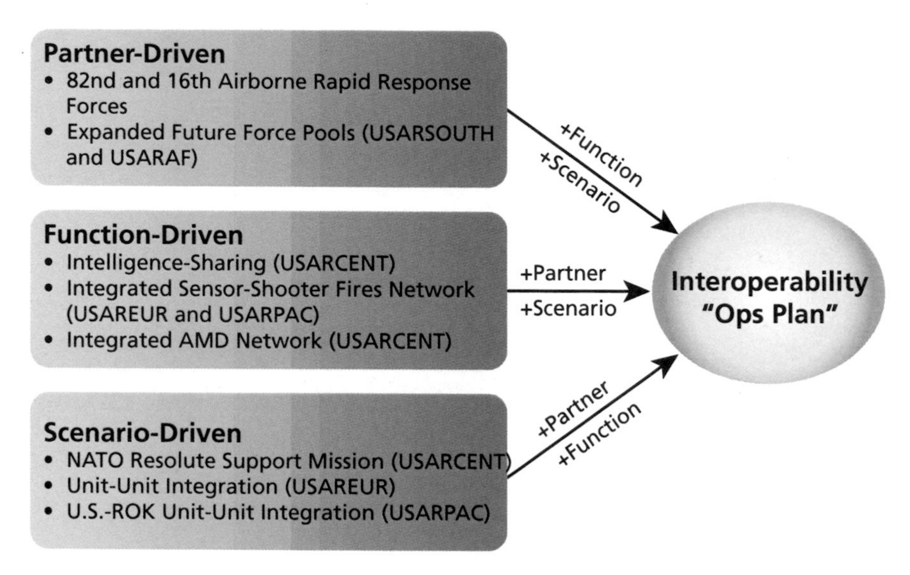

NOTE: USARSOUTH = U.S. Army South; USARCENT = U.S. Army Central; USAREUR = U.S. Army Europe; USARPAC = U.S. Army Pacific; AMD = air and missile defense.

82nd and 16th Airborne Rapid Response Forces

Ongoing cooperation between the U.S. 82nd Airborne Division and the UK 16th Airborne Brigade is one example of multinational interoperability built from an existing special relationship between partners. As a historically close ally with an overlapping language, culture, and military structure, the United Kingdom is a natural choice for interoperability. Furthermore, both countries' airborne forces maintain overlapping capabilities, equipment, and mission sets, making each an attractive interoperability candidate to the other.

Partnering U.S. and UK airborne forces provides a complementary set of functional capabilities and missions that, with enough interoperability, offers the potential of enhanced rapid response capabilities for global contingencies. Specifically, such joint response capabilities might reduce the resource demands on both partners by enabling joint U.S. and UK airborne forces to consolidate greater capabilities on a single contingency—or cooperatively respond to multiple contingencies—without spreading either states' forces too thin.

In this way, interoperability between the 82nd Airborne Division and 16th Airborne Brigade, though initially partner-driven, significantly affects functional and scenario contributions to multinational operations. Perhaps there is no better example of this partner-driven consolidation of resources on scenario and functional demands than both states' contributions to NATO's Very High-Readiness Joint Task Force (VJTF), NATO's rapid-response deterrence force.[1] By being more interoperable, U.S. and UK airborne forces can jointly contribute to NATO VJTF missions, which further improves both countries' readiness and responsiveness through shared responsibilities.

However, this requirement to effectively cooperate, consolidate resources, or do both in responding to global demands necessarily implies the need to at least share, if not sometimes integrate, capabilities. Jointly responding to certain scenarios means little if those multinational forces cannot also operate together in theater. Similarly, coordinating over global responses suggests a need to responsively communicate and plan for interdependent operations to maximize global responsiveness. Thus, though such relationships are partner-driven, the development of more-detailed functional and scenario requirements further drives how those multinational forces plan for and build interoperability.

Expanded Future Force Pools (USARSOUTH and USARAF)

In other cases, interoperability is driven not by existing special relationships between U.S. and partner forces but rather by the acknowledgement that building interoperability with certain partners might benefit future U.S. operations by expanding force pools from which to draw on during a conflict. This is especially true in instances of capable U.S. partners that are not necessarily needed for specific contingencies but can broadly contribute to U.S. global strategic aims, or partners

[1] Christine Wormuth, "How is DOD Responding to Emerging Security Challenges in Europe?" testimony before the House Armed Services Committee, Washington, D.C., February 25, 2015; Daniel Wasserby, "NATO 'Spearhead' Force to Take Shape by February 2015," *International Defence Review*, October 7, 2014.

that—with some enabling—can provide important capabilities to U.S. forces abroad. In our discussions across the ASCCs, we noted several prominent instances of U.S. Army units using SC and building partner capacity activities to develop partner nations' capabilities, thereby expanding future force pools and enabling potential future interoperability with key regional partners.

Two prominent examples of enabling partners to expand future force pools can be found in South America and Africa. U.S. Army forces have repeatedly worked with ground forces in El Salvador and Brazil to improve training, professionalism, and effectiveness. Such activities have paid dividends, given that forces from both countries deployed to Africa as part of United Nations (UN) peacekeeping missions, thus potentially undertaking the burden of maintaining international peace instead of U.S. forces needing to perform that duty.[2] Although these missions were not the original intent of U.S. SC activities, their utility in furthering U.S. global aims while reducing demands on U.S. resources highlights a potential avenue for focusing future U.S. SC activities on specific benefits from continued interoperability.

Similarly, U.S. Army forces operating with key partners in the Lake Chad Basin in north-central Africa developed a low-cost training program to develop partner countries' counter–improvised explosive device capabilities, centered around developing shared processes and training to U.S. and international standards of effectiveness. Capacity-building activities not only improve the capabilities of those partners for their own missions but also can provide future U.S. and UN coalitions with more potential participants who can bring specialized capabilities.[3] Again, those partner forces serve to reduce demands on U.S. military forces operating abroad by enhancing U.S.-based capabilities in specific missions with partner-specific capabilities. While those activities were originally undertaken with the goal of building partner capacity, those partners' abilities to then plug into larger U.S. and UN operations highlights opportunities to further focus SC activities.

---

[2] U.S. Army South personnel, interview with the authors, San Antonio, Tex., March 2018.

[3] U.S. Army Africa personnel, 2018.

Both of those benefits—enabling partners to enhance or replace U.S. forces abroad—appear in U.S. operations with regional partners against the terrorist group al-Shabaab in Somalia. As noted in the 2015 U.S. Africa Command Posture Statement,

> In the past year, with advice and assistance from U.S. forces, African Union forces improved their operational planning, demonstrated increased proficiency on the battlefield, and gained significant territory from al-Shabaab. During Operation INDIAN OCEAN, African Union forces liberated key terrain from al-Shabaab's control and disrupted the group's training, operations, and revenue generation. The African Union Mission in Somalia, United Nations, and East African partners improved their coordination in planning for offensive and stability operations.[4]

That is, U.S. SC assistance with African Union partners enabled those forces to more effectively operate to degrade key U.S. threats without relying heavily on U.S. forces.

Expanding future force pools and setting the stage for greater future interoperability with key partners provides several benefits. In the short term, such activities maintain U.S. access to partners through repeated contact, which both USARSOUTH and USARAF referenced as a key benefit in an era of increasing global competition with adversaries. Such activities can also reduce the costs and burdens on U.S. Army forces if, through U.S. capacity-building activities, those partners deploy for missions instead of U.S. troops. In the longer term, these activities can help the United States shape partners' capabilities as they develop, potentially helping to fill niche capability gaps or provide the ability for future coalitions to leverage specialized capabilities of partner countries. Therefore, while the benefits of SC and building partner capacity activities may not be immediately revealed at operational or tactical levels, such activities can provide geopolitical benefits in the short term and can set the stage for promoting such tactical-level interoperability in future scenarios.

---

[4] U.S. Africa Command, *United States Africa Command 2015 Posture Statement*, March 17, 2015, p. 8.

Like the relationship between the 82nd Airborne Division and the 16th Airborne Brigade, the cases from South America and Africa highlight how improvements in interoperability with key partners can reduce resource demands on U.S. forces in responding to certain demands or missions. However, in contrast to the U.S.-UK relationship, these benefits result from a U.S. enabling of partner capabilities, either through training, advising, or equipping.

Examples of Function-Driven Interoperability

Other times, interoperability is discussed in terms of functional needs and enabling interoperability to perform a certain warfighting function—such as joint fire or command and control—with multinational partners. That is, U.S. military forces identify the need for greater capability in some warfighting function in a region, and then work with specific partners to develop that capability. Interoperability built toward one or a few functions could be used in specific planning or operational scenarios, but the need for functional interoperability often spans multiple regions and threats. After identifying which functional capabilities are broadly required, U.S. forces then work with key regional partners to build those capabilities for application in specific missions or against specific threats.

Intelligence-Sharing (USARCENT)

Better intelligence-sharing with partners is often a key requirement to enable U.S. operations, and greater intelligence-based interoperability with partners is a joint force imperative. In many cases, regional partners are the best source of operational intelligence to enable U.S. operations because their language and cultural understanding and proximity to threat networks often affords them access that U.S. personnel do not have.[5] As a result, better intelligence-sharing with partners constitutes an increase in capabilities that U.S. forces cannot necessarily overcome

[5] Jason Welch, "Intel Workshop Combines Coalition and Iraqi Experiences," Defense Visual Information Distribution Service, August 13, 2018.

in isolation, thus potentially filling important capability gaps in U.S. operations.[6]

Indeed, as the 2018 U.S. Central Command posture statement points out about the importance of increasing intelligence-based interoperability to support counter-terrorism operations,

> the lack of national-level intelligence sharing agreements often hinders the timely and comprehensive communication of information. Our classified networks are largely unavailable to our partner nations and inhibit our ability to integrate operations, often requiring costly and labor-intensive solutions to overcome. However, utilizing a coalition-centric approach necessitates a paradigm shift and a deliberate acceptance of risk in order to foster an environment of reciprocal information sharing. We have an opportunity to sustain momentum in the global campaign against ISIS [Islamic State in Iraq and Syria] and other VEOs [violent extremist organizations] while continuing to refine the whole-of-coalition approach. Opposition to violent extremism provides unique alignment of national interests and can increase trust, understanding, and cooperation on other critical issues.[7]

While the need for better intelligence networks is not tied to any specific threat or partner, the 2018 U.S. Central Command posture statement does highlight the technical and procedural requirements to be able to exchange information quickly and securely. This suggests the need to focus on methods and equipment necessary to enable high-level sharing with partners focused on seamlessly enabling the key function of intelligence-sharing when necessary with key partners.

---

[6] Chad C. Serena, Isaac R. Porche III, Joel B. Predd, Jan Osburg, and Brad Lossing, *Lessons Learned from the Afghan Mission Network: Developing a Coalition Contingency Network*, Santa Monica, Calif.: RAND Corporation, RR 302-A, 2014.

[7] Joseph L. Votel, "The Posture of U.S. Central Command—Terrorism and Iran: Defense Challenges in the Middle East," statement before the House Armed Services Committee, Washington, D.C., February 27, 2018, p. 43.

Integrated Sensor-Shooter Fire Network (USAREUR and USARPAC)

Initiatives to advance cooperation on fire with regional partners similarly highlights the need for high-level integration of U.S. and partner capabilities to meet functional demands. A robust ability to call for joint fire is necessary for most high-end warfighting scenarios, and the ability to integrate fire with multinational partners can increase the density and robustness of fire networks among high-end peers. Thus, the United States spends considerable effort building integrated fire networks with high-end partners both in Europe and in the Pacific to counter and deter regional adversaries.[8]

For instance, during Dynamic Front 18, a multinational training exercise, Fire Support Teams of the 2nd Cavalry Regiment served as observers for fire missions prosecuted by the 1st Artillery Brigade of the British Royal Army.[9] This entailed digital transmission of firing data through sophisticated networks interconnecting the forces. Similarly, the USARPAC Multi-Domain Task Force Concept, predicated on integrated fire capabilities, is beginning to operate with multinational partners in training environments, ostensibly toward similarly seamless integration of capabilities.[10]

Integrated Air and Missile Defense Network (USARCENT)

Another example of such high-end shared capabilities to serve a functional purpose is the development of a regional missile defense network with Middle East partners. Since President Bill Clinton's administration, the United States has advocated the development of an integrated regional-defense system that leverages the national capabilities of the Gulf Cooperation Council states and Egypt. In December 2008, the integrated ballistic missile defense project in the Middle East was reportedly addressed at a conference of the International Institute for Strategic Studies in Bahrain; then–Secretary of Defense Robert Gates

---

[8] U.S. Army Europe personnel, interview with the authors, Wiesbaden, Germany, April 2018; U.S. Army Pacific personnel, interview with the authors, Honolulu, Hawaii, May 2018.

[9] James Anderson, "Saber Squadron Improves Sensor-to-Shooter Fires Interoperability," Defense Visual Information Distribution Service, April 9, 2018.

[10] Sean Kimmons, "Multi-Domain Task Force Set to Lead Pacific Pathways Rotation in First Overseas Test," Army News Service, June 15, 2018.

explained that the United States was "working both on a bilateral and a multilateral basis in the Gulf to establish the same kind of regional missile defense that would protect our facilities out there as well as our friends and allies."[11] In this case, a broad missile-defense network made up of regional partners could help counter potential threats to regional assets from Iranian ballistic missiles.

Like the need for high-end sharing of systems and procedures to enable information-sharing between partners, each of these cases relies heavily on the ability to tactically share partners' capabilities through similar technical systems, shared operational procedures, and intensive training. The benefits of that tactical sharing are significant. By enabling such multinational functionality, the United States can reduce regional demands for key capabilities (e.g., missile defense systems to singularly guard allies and installations) while also leveraging partner capabilities to fill or enhance important capability gaps in regional U.S. posture.

Examples of Scenario-Driven Interoperability

Interoperability planned and built to meet the needs of a scenario is relatively rare. It entails generating requirements for partners and allies to fight alongside the United States to win against specific adversaries. Often, the details of the scenario outline the key functions needed to win and the key partners involved in the fight. Such scenario-driven demands could include partners coordinating resources and integrating operations against a common adversary or sharing capabilities across multiple warfighting functions. Larger scenarios could entail a significant chance of the United States taking operational command and control of partner nations' units or vice versa. Like function-driven interoperability, such scenario-driven interoperability often requires intensive sharing of capabilities to meet scenario demands.

---

[11] Claude Salhani, "U.S. Wants Gulf to Buy into Missile Defense System," *Middle East Times*, December 15, 2008; Robert Gates and James Cartwright, "DoD News Briefing with Secretary Gates and Gen. Cartwright from the Pentagon," U.S. Department of Defense, September 17, 2009.

NATO Resolute Support Mission (USARCENT)

Coalition operations in Afghanistan, including Operation Freedom's Sentinel and the NATO Resolute Support Mission, are perhaps the best ongoing example of scenario-driven interoperability. The NATO Resolute Support Mission was launched in January 2015, following the completion of the mission of the International Security Assistance Force in December 2014, when responsibility for security in Afghanistan was transferred to the Afghan national defense and security forces.

From a strategic perspective, the multinational dimension of NATO Resolute Support Mission—roughly 39,000 troops from 39 NATO and partner countries—significantly increases the international legitimacy of the ongoing U.S.-led Afghanistan stabilization mission.[12] Operating effectively with international partners has allowed U.S. forces to leverage NATO capabilities and reduce the overall U.S. resource burden for ongoing operations in Afghanistan while also leveraging partners' specialized capabilities, such as the United Kingdom's specialization in governance and development.[13] With many capable partners and a robust division of responsibility throughout the country, this interoperability is primarily facilitated through the simple enabling of partner operations through the existing NATO alliance framework.

However, in a few select cases, simple coordination of efforts might not be sufficient to meet scenario demands, and U.S. units may need to develop interoperability with capable partners to enable combined operations across a wide range of warfighting functions.[14] Such instances are usually the result of repeated and dedicated interactions with highly capable partners facing a significant and concerted threat.

Unit-Unit Integration (USAREUR)

In the USAREUR area of operation, U.S. Army units have repeatedly shown the ability to operate and effectively maneuver under the control of, and with control over, maneuver units of tier-one NATO

---

[12] NATO Resolute Support Mission Afghanistan, "About Us: Mission," webpage, undated.

[13] Alexander Powell, Larry Lewis, Catherine Norman, and Jerry Meyerle, *Summary Report: U.S.-UK Integration in Helmand*, Washington, D.C.: CNA, February 2016.

[14] Serena et al., 2014.

partner nations, including those of Canada, France, and the United Kingdom, during operations while integrating multiple warfighting functions. Such close unit integration requires significant investments in both time and technical capabilities. For instance, in addition to needing compatible command and control, intelligence, communications, and logistics systems to support combined operations, combined units must work under common standards and processes, such that all units can speak a common operational language when operating together, even if their national languages differ; for example, multinational units must be able to call for fire or logistics support along a shared concept of support to effectively enable a combined maneuver.

The ability to field combined units that can maneuver and operate together provides strategic and operational benefits. Strategically, tactical unit integration is a significant force and readiness multiplier, and the ability to interoperate with partners increases deterrence by presenting a stronger, more capable, and more responsive force in the face of adversary aggression. Fielding combined forces, even those that include majority-U.S. forces, can also help fill U.S. readiness gaps and ensure U.S. ability to meet NATO obligations while reducing resource demands on U.S. forces. Multinational forces operating together also increase the international legitimacy of coalition operations, whether it be multinational forces fighting together in the Gulf War or NATO forces operating together in defense of the Baltic states against Russian aggression.[15] Operationally, being able to effectively maneuver and fight together, and better understanding how partners operate through repeated interactions, can increase the safety of forces involved through a shared understanding of the battlespace.

U.S.-ROK Unit-Unit Integration (USARPAC)

The example of tactical unit integration with the largest scope is the ongoing U.S.-ROK operational relationship. As part of their ongoing alliance, U.S. and ROK units—spanning every functional capability from planning to fire to maneuver to aviation—train and oper-

[15] Gordon and Trainor, 1995.

ate together as a combined force.[16] Importantly, this interoperability occurs at multiple echelons from battalions and brigades to divisions and corps-level coordination. Perhaps nowhere is this unit coordination more evident than in the combined U.S.-ROK staff of the 2nd Infantry Division, which is led by a combined U.S.-ROK staff and, in the case of combined operations, could command both U.S. and ROK maneuver units. The 2nd Infantry Division is notable for facilitating combined U.S.-ROK planning and operations because of the trust and shared unity of purpose it fosters.[17]

Over time, as ROK military capabilities have increased, U.S.-ROK interoperability has similarly increased and deepened, allowing the United States to transition from the defense of South Korea to the support of an ROK-led defense of its territory. While supporting what is primarily a partner-led mission, the U.S. ability to train, exercise, and operate with ROK Army units serves as a significant deterrent to North Korean aggression. At an operational level, the ability to operate together increases unit safety through better coordination during operations. Perhaps most important, however, is the way that such combined operations fill capability gaps and allow both U.S. and ROK militaries to leverage specialized capabilities during operations. This point is perhaps best displayed in the combined U.S.-ROK counter–weapons of mass destruction mission which, by the nuclear non-proliferation treaty, must be completed by U.S., rather than ROK, forces.[18]

Requiring greater integration of capabilities, these three cases also highlight the need for greater sharing of capabilities as the functional integration of U.S. and partner units increases. That is, while the U.S. enabling of partner capabilities is sufficient to effectively produce the high-level coordination needed in Afghanistan, the tactical integration of capabilities in European and Korean scenarios often entails more-significant sharing of operational systems between partners in those

[16] Eighth United States Army and 2nd Infantry Division personnel, interview with the authors, Seoul, Republic of Korea, February 2018.

[17] Eighth United States Army and 2nd Infantry Division personnel, 2018.

[18] Center for Army Lessons Learned, 2015.

regions. Overall, this suggests that, as the capabilities shared between partners become increasingly tactical, the level of integration between partners must necessarily increase to produce the desired outcomes from interoperability.

Summary

The preceding examples highlight the myriad strategic and operational benefits interoperability might bring to U.S. Army forces, provided that those benefits match the type and level of investment undertaken by U.S. forces. High-level strategic gains and general benefits from broad resource-sharing with partners are perhaps best realized through the enabling of partner forces. In contrast, relying on interoperability with partner forces to fill critical capability gaps and provide effective tactical solutions to warfighting problems requires a significant sharing of capabilities at a much lower operational level. The key to realizing these benefits across partners lies in fully understanding what benefits U.S. forces are seeking in any given situation and how to most effectively approach building interoperability toward those specific benefits. Being explicit about answering the *who*, the *what*, and the *what for* questions for interoperability planning will help build effective interoperability.

CHAPTER FOUR

A Preliminary Examination of the Risks and Resourcing Demands of Interoperability

While the previous chapters explored what benefits might be driving interoperability investments, this chapter examines some of the risks and resourcing demands involved.[1] We first examine some of the interoperability risks previous studies have identified. We then discuss interoperability costs that Army stakeholders have observed based on recent Army interoperability activities.

Interoperability Can Entail Risks

Relying on partners can expose the United States to strategic and operational risks. As Nora Bensahel cautions, multilateral military operations often entail trade-offs at both the strategic and operational levels.[2] We outline some of these risks in the following paragraphs.

Disagreement over strategic objectives. Working with partners requires reaching an agreement on the political goals of the operation. This can be difficult when different partners have competing preferences over the objectives of multinational actions. Consider an example in which one partner wants to undertake a military intervention and

[1] Understanding the value of interoperability is necessarily a net concept in which the costs and benefits are diverse and often not easily quantified. The evaluation framework developed in O'Mahony et al., 2018, provides a potential methodological template for evaluating interoperability investments.

[2] Bensahel, 2007.

the other partner does not. The first partner faces the risk of abandonment, and the second partner decides not to intervene. To the extent that the countries' forces rely on each other's capabilities, the intervening partner might have capability gaps when operating alone.[3] In contrast, the second partner faces the risk of entanglement in operations that the country would not have chosen to get involved in on its own.[4]

Disagreement over how to accomplish strategic objectives. Even when partners agree on strategic objectives, they may not agree on what actions to take to accomplish their shared objectives. For example, as Raphael Cohen and Andrew Radin discuss, although both the United States and Baltic countries share a strategic objective to deter potential Russian hostile actions in the Baltics, they diverge in their preferred strategies for doing so. [5] These disagreements can affect not only what activities are undertaken during operations but also how partners prepare and train for operations. For example, negotiations related to how many training activities, for what capabilities, and at what facilities have long been monitored—and objected to—by potential adversaries. The U.S. decision to not hold the 2018 Freedom Guardian exercises with South Korea is one such case.

Strategic and operational constraints because of national caveats. National caveats limit the capabilities that a partner can deploy. These allow partner governments to tailor their participation in military operations to activities that are politically palatable. If these caveats are not known in advance, they can represent a significant vulnerability to military operations. David Auerswald and Stephen Saideman document the national caveats used by NATO forces in Afghanistan and

---

[3] This is of particular importance when relying on partners for critical niche capabilities. As a result, Army planners tend to be risk-averse when planning activities, often not incorporating partner capabilities into their plans.

[4] David A. Lake, *Entangling Relations: American Foreign Policy in Its Century*, Princeton, N.J.: Princeton University Press, 1999.

[5] Raphael Cohen and Andrew Radin examine how competing preferences within NATO shape planning to counter potentially hostile actions by Russia in Europe. Raphael S. Cohen and Andrew Radin, *Russia's Hostile Measures in Europe: Understanding the Threat*, Santa Monica, Calif.: RAND Corporation, RR-1793-A, 2019, pp. 152–153.

find that these caveats reduced the effectiveness of NATO operations and led to resentment across NATO forces.[6]

Operational constraints because of the need for compatible doctrine, processes, and materiel. Unless carefully planned throughout development and acquisition, the Army's objectives for interoperability and modernization can run counter to each other. Advances in U.S. doctrine, processes, and materiel can outstrip those of partners. As the United States modernizes its systems, it faces a set of alternatives: Maintain older systems that are interoperable with partners, develop modernization campaign plans with key partners, provide new systems to key partners, or undertake more-limited interoperability in the areas where the United States prioritizes its modernization.

Operational constraints because of poor agility in command and control processes. Historically, coalition operations have been constrained by cumbersome multinational command and control procedures.[7] Patricia Weitsman argues that it can be difficult to transform a peacetime command and control structure to a wartime footing.[8] The lower the echelon at which consensual partner command and control decisions need to be made, the less agile operational and tactical decision-making will be.

---

6 David P. Auerswald and Stephen M. Saideman, *NATO in Afghanistan: Fighting Together, Fighting Alone*, Princeton, N.J.: Princeton University Press, 2014.

7 Nora Bensahel assesses the adverse impact of multiple chains of command facing ground forces in the Gulf War, Somalia, and Bosnia. See Bensahel, 2007. Robert Tripp and colleagues document the difficulties multiple national chains of command caused in Operation Enduring Freedom. Robert S. Tripp, Kristin F. Lynch, John G. Drew, and Edward W. Chan, *Supporting Air and Space Expeditionary Forces: Lessons from Operation Enduring Freedom*, Santa Monica, Calif.: RAND Corporation, MR-1819-AF, 2004.

8 Patricia A. Weitsman, "Wartime Alliances Versus Coalition Warfare: How Institutional Structure Matters in the Multilateral Prosecution of Wars," *Strategic Studies Quarterly*, Vol. 4, No. 2, 2010.

U.S. Army Stakeholders Identified Interoperability Costs

The U.S. Army and its partner ground forces allocate and expend resources—chiefly, time and money—in pursuit of the potential benefits that multinational interoperability offers. Army organizations commit staff and leadership time, effort, and focus to envision, plan, execute, and administer interoperability-related initiatives and events. In cases where interoperability is a command priority, this commitment is nontrivial. When resources are not explicitly assigned to support interoperability, these demands force sometimes difficult trade-offs—even when the overall level of resources demanded is not high. Collateral duties or special positions related to multinational interoperability appear to be a norm. Examples related to enhancing human, procedural, and technical interoperability show the activities needed to achieve such interoperability and the costs of such activities (Table 4.1).

Most of the command staff we interviewed emphasized that the costs of achieving interoperability entail the leadership time, effort, and focus required to advance interoperability. Each of the activities in Table 4.1 requires some level of funding: for travel to a partner ground force's event, for interoperability-related incremental expenses during bilateral training, and for software vendors that offer common operating picture solutions. However, relative to unit operating budgets, planners and resource managers we interviewed indicated that these expenses, when planned for, are manageable and often shared with partner countries. At the HQDA and Army command levels, the time-money dynamic is similar. Funding does not appear to be a significant barrier when requested and programmed in advance.

Although the costs of achieving interoperability are often the costs associated with time spent and opportunity costs lost, achieving interoperability does incur direct financial burdens; however, such direct financial burdens are difficult to discern. We are unaware of any operational commands that holistically track spending for multinational interoperability, and ASCC activities often have multiple, layered objectives for operations and exercises. In our site visits and interviews and in our review of exercise after-action reviews, we identified

Table 4.1
Costs to Achieve Different Types of Interoperability

Type	Example	Activities Needed to Enhance Interoperability	Costs of Activities
Human	A forward deployed unit embraces Army best practices for establishing unit-to-unit and commander-to-commander interpersonal relationships with a partner nation ground unit that is expected to operate with during wartime.[a]	The commander attends the partner unit's change of command ceremonies, promotions, and national holiday celebrations. Meanwhile, select officers and senior noncommissioned officers deliberately reach out to their counterparts to build rapport beyond what can be established during bilateral training events.	Activities are likely to incur an opportunity cost in the form of less planning time, training hours, oversight over junior soldiers, or rest and recovery.
Procedural	An Army division maintains a habitual relationship with an allied brigade, and it may be called on to conduct integrated crisis response operations.[b]	A core group of officers reporting directly to a division deputy commander spends most of its time on administrative, procedural, logistical, and security matters required to integrate the partner brigade into exercises and potential operations.	Planning and coordinating for highly targeted multinational interoperability (i.e., unit integration) may result in less attention to the division's other priorities.[c]
Technical	During site visits, each military unit relayed that it has an interoperability-driven need for a common operational picture (COP) that includes allied forces, ideally updated with minimal latency and pushed down to a specified tactical echelon.[d]	Several units are in the process of developing requirements for such a COP, exploring software options, and coordinating with partner nation ground forces.	This process may take several years of intensive work by planners, operators, and technical staff, with the commander (the ultimate consumer of a COP) needing to provide iterative guidance along the way.

[a] One example of the emphasis placed on social interaction with partner units is found in a Center for Army Lessons Learned handbook on multinational interoperability in Korea: "It cannot be over emphasized that a lack of depth with personal relationships leads to lack of planning effectiveness that hurts interoperability. Conversely, effective social interaction can overcome rank and other hierarchical factors and build effective interoperable teams" (Center for Army Lessons Learned, 2015, p. 148).
[b] The U.S. Army maintains this type of relationship with select partner country armies through memorandums of understanding. HQDA, G-357 Strategy staff, interview with the authors, Washington, D.C., May 22, 2018.
[c] 82nd Airborne Division staff, interview with the authors, Fort Bragg, N.C., March 13, 2018.
[d] ASCC and theater Army staff, interviews with the authors, various locations, Spring 2018.

the following two ways in which interoperability activities tended to be funded:

- **Interoperability-unique funding** is for the primary purpose of exchanging services to operate effectively together with partner nations
- **Interoperability-enhancing funding** is spent for other purposes but has a discernable (or intended) impact on operating effectively with partner nations.

Table 4.2 provides a nonexhaustive list of monetary costs associated with multinational interoperability, divided by interoperability-unique or interoperability-enhancing funding columns. The rows represent ten groups of Army SC activities identified in previous RAND research.[9] Empty cells indicate that we did not identify funding within the activity group.

As shown by empty cells in the interoperability-unique funding column, few organizations or units exist for the primary purpose of fostering multinational interoperability. The most active and visible example of a purpose-built entity at the ASCC level is the 30-soldier Digital Liaison Detachment (DLD) construct, which provides "a digital liaison capability to Army units (theater army, corps, and division headquarters) for connectivity with allied and multinational force units and other U.S. Services."[10] Army leaders or resource managers interested in understanding the scope of spending for multinational interoperability could begin by assembling a complete list of similar organizations and proceed to identify the fully burdened cost (i.e., inclusive of personnel, equipment, training, facilities) of the Army maintaining such structure. Other interoperability-unique costs include initiatives to establish secure networks or networking environ-

---

[9] See Pernin et al., 2019.

[10] HQDA, *Digital Liaison Detachment*, Washington, D.C.: Army Publishing Directorate, ATP 3-94.1, December 2017. These detachments can facilitate such functions as mission command, combined fire, and information-sharing. Two DLDs are currently assigned to provide connectivity between the ROK Army and the 8th Army Headquarters, and a third belongs to USARCENT.

Table 4.2
Select Monetary Costs for Multinational Interoperability

Activity Group	Interoperability-Unique	Interoperability-Enhancing
Training and exercises	—	• Warfighter exercises • Joint Multinational Readiness Center
Staff exchanges	—	• Army staff talks • Military Personnel Exchange Program positions
Consultations and information exchanges	• Interoperability boards (e.g., ABCANZ, NATO)	• Senior leader travel
Education	—	• Courses for attachés, security cooperation officers, advisers, foreign area officers, foreign disclosure officers, and ASCC planners (e.g., Security Cooperation Planners Course)
Research, development, test, and evaluation (RDT&E)	• Mission Partner Environment and network-related initiatives • Coalition interoperability assurance and validation • Compatible radios	• Adjustments to requirements for existing or future systems • RDT&E involving international partners
Armaments and arms control	—	• Deputy Assistant Secretary of the Army for Defense Exports and Cooperation[a]
Unit-to-unit relationships	• Security for a multinational facility	• Commander conferences
Equipment transfers	• Cryptographic information or devices	• Foreign weapon sales and transfers[a] • United States Army Security Assistance Command[a] • United States Military Training Mission in Saudi Arabia[a]
Liaison officers	• DLDs • Formal and ad-hoc unit-to-unit liaison officers	• Foreign liaison officers
Multinational operations	• Narrowly focused operations	• Most operations • Security Force Assistance Brigade deployments

[a] Funded entirely or partially by Foreign Military Sales or a similar mechanism.

ments suitable for multinational operations, and HQDA-facilitated forums, such as the American, British, Canadian, Australian, and New Zealand (ABCANZ) Armies' Program or participation in the NATO Military Committee Land Standardization Board.

As the second column shows, interoperability-enhancing costs are more widespread. The ASCCs typically conduct training events, exercises, and multinational operations for purposes beyond multinational interoperability—in other words, the activities would still occur in the absence of interoperability-related objectives. Most planners and resource managers we interviewed indicated that the incremental cost of adding interoperability objectives into events represents a small fraction of the activity's overall cost.

Summary

The unclear and often unstated benefits of interoperability are similarly manifested in how it is resourced. Bottom-up efforts to bring forces together in the hopes of building interoperability are levied on existing processes and activities, often with additional costs and in competition with other activities. Thus, tactical units bear the brunt of the interoperability demands. In this case, interoperability is used as a justification without clear value. Moving toward the future, it will be necessary to have a clearer understanding of the benefits interoperability brings, and mechanisms for properly inserting those requirements within units and commands.

CHAPTER FIVE

Potential Next Steps

The benefits ascribed to interoperability vary widely and are not always well articulated or matched to the types of interoperability developed. The United States works closely with foreign militaries to close capability gaps, shape the strategic environment, and reduce resource demands when meeting national interests. These potential benefits combine in ways that depend on the nature of the engagement: which partners and for what functions and scenarios the forces hope to be interoperable. These situations are dynamic; the environment today is not what it might be in the future.

From this look at the benefits that interoperability might bring and how such benefits are constructed in a cross-section of the Army's interoperability initiatives, we offer some potential next steps for the Army:

- Do not assume that high levels of interoperability are valuable, or even possible, with most partners. One size does not fit all.
- Define requirements for interoperability based on the benefits that interoperability can provide. Interoperability is a *means to some other end*—not an end in itself. We identified three overarching objectives for pursuing interoperability: shape the strategic environment, increase multinational capabilities, and reduce resource demands.
- Be specific about the benefit expected from any given investment to build interoperability. Develop metrics to evaluate interoperability outcomes. Track interoperability built in line with the specified benefits, and hold those involved accountable.

- Do not include interoperability as an objective in strategic documents (including country plans), unless the purpose of interoperability is clearly defined in the document in terms of end state and benefits.
- Examine U.S. Army processes for obstacles to building interoperability. Not knowing or articulating the benefit of interoperability is only one of many organizational obstacles to building interoperability. A few potential obstacles include securing resources, developing new doctrine for operating with partners, and developing training systems and readiness metrics for an interoperable force.
- Assign a proponent for interoperability. The proponent should be responsible for identifying and proposing fixes for institutional obstacles. Building and sustaining interoperability will require many higher-level commands to behave in sync. For instance, how programs are crafted in the Pentagon; how country strategies are set across combatant commands and HQDA; and how forces are trained at home station and abroad. The right institutional proponent, likely at the 4-star level, will need to corral such a group for broad benefits not accrued in any one command.
- Interoperability is a partnership. Work closely with partners to jointly develop both the potential values and development strategy for interoperability investments. The U.S. Army enjoys many high-level arenas for engaging with partner armies, such as the staff talks run through the Army's Deputy Chief of Staff for Operations (G-3/5/7). These and other forums will remain important to agenda-setting and communicating challenges and solutions on both sides.

Understanding the benefits is but one early and important step in moving the U.S. Army forward on interoperability. Building multinational interoperability brings costs and risks that will also need to be weighed as the Army competes against other capabilities in a resource-constrained environment. Aiming the institutions in the right direction, with the articulation of why interoperability needs to be developed, will be instrumental in eventually chasing down the benefits thereof and building a better force.

APPENDIX
Research Approach

We adopted a multipronged approach to identify what benefits interoperability can provide. We examined what has been said and written about interoperability and conducted interviews with a range of Army and DoD stakeholders.[1]

Literature Review

We identified interoperability benefits discussed in such venues as research studies, senior leader statements, national security strategy documents, bilateral and multilateral strategic vision statements, and Army operational guidance and lessons learned from past and ongoing operations. In the next section, we list some of our key sources.

Research Studies

The following are the studies we found most useful.

Center for Army Lessons Learned, "Commander's Guide to Multinational Interoperability," Fort Leavenworth, Kan.: U.S. Army Combined Arms Center, No. 15-17, 2015.

[1] It is important to note that this research approach was tailored to identify benefits that Army personnel expect or have experienced from multinational interoperability. It was not designed to evaluate either the impact of or costs associated with interoperability. Further research into the costs of interoperability will need to go beyond discussions of current activities and the costs observed by Army headquarters and operators.

Coleman, Katharina P., "The Legitimacy Audience Shapes the Coalition: Lessons from Afghanistan, 2001," *Journal of Intervention and Statebuilding,* Vol. 11, No. 3, 2017, pp. 339–358.

Ford, Thomas, John Colombi, Scott R. Graham, and David R. Jacques, "A Survey on Interoperability Measurement," *Proceedings of the 12th International Command and Control Research and Technology Symposium*, Newport, R.I., June 19–21, 2007.

Leeds, Brett Ashley, "Alliance Reliability in Times of War: Explaining State Decisions to Violate Treaties," *International Organization*, Vol. 57, No. 3, 2003, pp. 801–827.

Liska, George, *Nations in Alliance*, Baltimore, Md.: Johns Hopkins University Press, 1962.

Moroney, Jennifer D. P., David E. Thaler, and Joe Hogler, *Review of Security Cooperation Mechanisms Combatant Commands Utilize to Build Partner Capacity*, Santa Monica, Calif.: RAND Corporation, RR-413-OSD, 2013.

O'Mahony, Angela, Ilana Blum, Gabriela Armenta, Nicholas Burger, Joshua Mendelsohn, Michael J. McNerney, Steven W. Popper, Jefferson P. Marquis, and Thomas S. Szayna, *Assessing, Monitoring, and Evaluating Army Security Cooperation: A Framework for Implementation,* Santa Monica, Calif.: RAND Corporation, RR-2165-A, 2018.

O'Mahony, Angela, Thomas S. Szayna, Christopher G. Pernin, Laurinda L. Rohn, Derek Eaton, Elizabeth Bodine-Baron, Joshua, Mendelsohn, Osonde A. Osoba, Sherry Oehler, Katharina Ley Best, and Leila Bighash, *The Global Landpower Network: Recommendations for Strengthening Army Engagement*, Santa Monica, Calif.: RAND Corporation, RR-1813-A, 2017.

Pernin, Christopher G., Katharina Ley Best, Matthew E. Boyer, Jeremy M. Eckhause, John Gordon IV, Dan Madden, Katherine Pfrommer, Anthony D. Rosello, Michael Schwille, Michael Shurkin, and Jonathan P. Wong, *Enabling the Global Response Force: Access Strategies for the 82nd Airborne Division*, Santa Monica, Calif.: RAND Corporation, RR-1161-A, 2016.

Pernin, Christopher G., Jakub P. Hlavka, Matthew E. Boyer, John Gordon IV, Michael Lerario, Jan Osburg, Michael Shurkin, and Daniel C. Gibson, *Targeted Interoperability: A New Imperative for Multinational Operations*, Santa Monica, Calif.: RAND Corporation, RR-2075-A, 2019.

Pernin, Christopher G., Angela O'Mahony, Thomas S. Szayna, Derek Eaton, Katharina Ley Best, Elizabeth Bodine-Baron, Joshua Mendelsohn, and Osonde A. Osoba, "What Is the Global Landpower Network and What Value Might It Provide?" *Defense and Security Analysis*, Vol. 33, No. 3, 2017, pp. 209–222.

Schmitt, Olivier, *Allies That Count: Junior Partners in Coalition Warfare*, Washington, D.C.: Georgetown University Press, 2018.

Shlapak, David A., and Michael Johnson, *Reinforcing Deterrence on NATO's Eastern Flank: Wargaming the Defense of the Baltics*, Santa Monica, Calif.: RAND Corporation, RR-1253-A, 2016.

Weitsman, Patricia A., *Waging War: Alliances, Coalitions, and Institutions of Interstate Violence,* Stanford, Calif.: Stanford University Press, 2014.

National Security Strategy Documents

We examined national strategic documents, including the 2018 National Defense Strategy, 2014 National Military Strategy, and recent Quadrennial Defense Reviews. We also examined service- and theater-level country, campaign, and security cooperation plans.

Operational Guidance

We examined service and joint doctrine that included references to interoperability. The most relevant ones for our study were as follows:

Army Pamphlet 11-31, *Army Security Cooperation Handbook*, Washington, D.C.: Headquarters, Department of the Army, February 6, 2015.

Army Regulation 11-31, *Army Security Cooperation Policy*, Washington, D.C.: Headquarters, Department of the Army, March 21, 2013.

Army Regulation 34-1, *Multinational Force Interoperability*, Washington, D.C.: Headquarters, Department of the Army, July 10, 2015. (In addition to draft updates.)

Field Manual 3-07.1, *Security Force Assistance*, Washington, D.C.: Headquarters, Department of the Army, May 1, 2009.

Field Manual 3-22, *Army Support to Security Cooperation*, Washington, D.C.: Headquarters, Department of the Army, January 2013.

Joint Publication 1-02, *Department of Defense Dictionary of Military and Associated Terms*, Washington, D.C., 2010.

Joint Publication 3-16, *Multinational Operations*, Washington, D.C., 2013.

Stakeholder Discussions

Stakeholders

We met with more than 100 people for this study, focusing on personnel with backgrounds in operations, intelligence, plans, fires, sustainment, and communications. We interviewed personnel at each ASCC, some GCCs, HQDA, research offices, and operational units, including the following:

82nd Airborne

Center for Army Analysis

Center for Army Lessons Learned

Office of the Secretary of Defense, Policy

U.S. Army, Headquarters of the Department of the Army G-3/5/7

U.S. Army, Headquarters of the Department of the Army G-8

U.S. Army Africa

U.S. Army Central

U.S. Army Europe

U.S. Army North

U.S. Army Pacific

U.S. Army South

U.S. Central Command

Interview Protocol

We focused our discussions around the following interview protocol, which was shared with interviewees prior to our meetings.

> RAND Arroyo Center is conducting a project for HQDA/G3-SS, MG McPadden (POC: Col Joseph Fossey, DAMO-SSC, 703-692-8781) on the "Value of Interoperability to the Army." The objective of this project is to assess the costs, benefits, and risks of interoperability to the U.S. Army. As part of this project we are conducting interviews across the Geographic Combatant

Commands and the ASCCs to help determine the value of multinational (MN) interoperability for reaching overarching goals and objectives. To that end, we are respectfully requesting your assistance in supporting a RAND team to visit and conduct one-on-one interviews in the coming months.

We would appreciate help in setting up meetings to discuss the following questions with staff representatives from Operations, Intelligence, Plans, Fires, Sustainment, Communications, and other cells if possible:

Past Examples

1. What are recent examples where the Army has effectively built interoperability with MN partners? What did it take to get there?
2. What are recent examples where the Army has not been able to build interoperability with MN partners? What were the main challenges? Were these challenges discovered while attempting to build interoperability, or did they stop the Army from attempting to build MN interoperability?

Linking Values of Interoperability to Goals

3. What are your command objectives? Where do you think building MN interoperability falls in terms of command priorities? In what command objectives is MN interoperability a necessary condition for success? (Please provide feedback on RAND's current collection.)
4. RAND has outlined several value propositions for interoperability, taken from the literature and interviews. How do these link to your command's objectives?

5. In your war plans, what are the most important examples of demand for interoperability? Which specific functions and with which specific partners/units is it most important? During which phases (Phase 0/1, Phase 3, etc.) is it most important?
6. With which partners do you have ongoing efforts to build interoperability for a primary purpose other than supporting a particular war plan? What is that purpose?
7. In terms of priority scenarios in your GCC, where are the biggest vulnerabilities to not being interoperable with MN partners? Which partners and which functions are of most concern?
8. What are your expected costs of building the interoperability you need, both in terms of systems and training time? How are those investments traded against other investments your command could be making?
9. Where in your command is "standing" interoperability most important? Where can interoperability be delayed and built "on the fly" when needed by a specific scenario?

Ongoing Activities

10. What current activities build interoperability? Is there evidence that they contributed to actual interoperability that was valued, demanded, or used?

References

82nd Airborne Division staff, interview with the authors, Fort Bragg, N.C., March 13, 2018.

Anderson, James, "Saber Squadron Improves Sensor-to-Shooter Fires Interoperability," Defense Visual Information Distribution Service, April 9, 2018. As of September 12, 2019:
https://www.dvidshub.net/news/272249/saber-squadron-improves-sensor-shooter-fires-interoperability

Anderson, Mindy, "U.S. Army Africa 'Train the Trainers' in Ghana," U.S. Africa Command, July 1, 2011. As of October 29, 2018:
https://www.africom.mil/media-room/Article/8443/us-army-africa-train-the-trainers-in-ghana

Army Pamphlet 11-31, *Army Security Cooperation Handbook*, Washington, D.C.: Headquarters, Department of the Army, February 6, 2015. As of October 29, 2018:
https://armypubs.army.mil/epubs/DR_pubs/DR_a/pdf/web/p11_31.pdf

Army Regulation 11-31, *Army Security Cooperation Policy*, Washington, D.C.: Headquarters, Department of the Army, March 21, 2013. As of October 29, 2018:
https://fas.org/irp/doddir/army/ar11-31.pdf

ASCC and theater Army staff, interviews with the authors, various locations, Spring 2018.

Auerswald, David P., and Stephen M. Saideman, *NATO in Afghanistan: Fighting Together, Fighting Alone*, Princeton, N.J.: Princeton University Press, 2014.

Barcelo, Nastasia, "Uruguay and the U.S. Train to Enhance Peacekeeping Missions," *Diálogo Digital Military Magazine*, August 14, 2015. As of October 29, 2018:
https://dialogo-americas.com/en/articles/uruguay-and-us-train-enhance-peacekeeping-missions

Bensahel, Nora, "International Alliances and Military Effectiveness: Fighting Alongside Allies and Partners," in Risa Brooks and Elizabeth Stanley, eds., *Creating Military Power: The Sources of Military Effectiveness*, Stanford, Calif.: Stanford University Press, 2007, pp. 186–206.

Center for Army Lessons Learned, "Commander's Guide to Multinational Interoperability," Fort Leavenworth, Kan.: U.S. Army Combined Arms Center, No. 15-17, 2015.

Cohen, Raphael S., and Andrew Radin, *Russia's Hostile Measures: Understanding the Threat*, Santa Monica, Calif.: RAND Corporation, RR-1793-A, 2019. As of August 1, 2019:
https://www.rand.org/pubs/research_reports/RR1793.html

Coleman, Katharina P., "The Legitimacy Audience Shapes the Coalition: Lessons from Afghanistan, 2001," *Journal of Intervention and Statebuilding*, Vol. 11, No. 3, 2017, pp. 339–358.

DoD—*See* U.S. Department of Defense.

Eighth United States Army and 2nd Infantry Division personnel, interview with the authors, Seoul, Republic of Korea, February 2018.

Ford, Thomas, John Colombi, Scott R. Graham, and David R. Jacques, "A Survey on Interoperability Measurement," *Proceedings of the 12th International Command and Control Research and Technology Symposium*, Newport, R.I., June 19–21, 2007.

Gates, Robert, and James Cartwright, "DoD News Briefing with Secretary Gates and Gen. Cartwright from the Pentagon," U.S. Department of Defense, September 17, 2009. As of November 12, 2009:
https://archive.defense.gov/transcripts/transcript.aspx?transcriptid=4479

Gordon, Michael R., and Bernard E. Trainor, *The Generals' War: The Inside Story of the Conflict in the Gulf*, Boston, Mass.: Little Brown and Co., 1995.

Headquarters, Department of the Army, *Digital Liaison Detachment*, Washington, D.C.: Army Publishing Directorate, ATP 3-94.1, December 2017.

Headquarters, Department of the Army, G-357 Strategy staff, interview with the authors, Washington, D.C., May 22, 2018.

HQDA—*See* Headquarters, Department of the Army.

Joint Publication 1-02, *DoD Dictionary of Military and Associated Terms*, JP 1-02, Washington, D.C., January 10, 2000.

Joint Publication 1-02, *DoD Dictionary of Military and Associated Terms*, Washington, D.C, 2010.

Kimmons, Sean, "Multi-Domain Task Force Set to Lead Pacific Pathways Rotation in First Overseas Test," Army News Service, June 15, 2018.

Lake, David A., *Entangling Relations: American Foreign Policy in Its Century*, Princeton, N.J.: Princeton University Press, 1999.

Leeds, Brett Ashley, "Alliance Reliability in Times of War: Explaining State Decisions to Violate Treaties," *International Organization*, Vol. 57, No. 3, Autumn 2003, pp. 801–827.

Liska, George, *Nations in Alliance*, Baltimore, Md.: Johns Hopkins University Press, 1962.

Moroney, Jennifer D. P., David E. Thaler, and Joe Hogler, *Review of Security Cooperation Mechanisms Combatant Commands Utilize to Build Partner Capacity*, Santa Monica, Calif.: RAND Corporation, RR-413-OSD, 2013. As of August 1, 2019:
https://www.rand.org/pubs/research_reports/RR413.html

NATO Resolute Support Mission Afghanistan, "About Us: Mission," webpage, undated. As of August 1, 2019:
https://rs.nato.int/about-us/mission.aspx

O'Mahony, Angela, Ilana Blum, Gabriela Armenta, Nicholas Burger, Joshua Mendelsohn, Michael J. McNerney, Steven W. Popper, Jefferson P. Marquis, and Thomas S. Szayna, *Assessing, Monitoring, and Evaluating Army Security Cooperation: A Framework for Implementation*, Santa Monica, Calif.: RAND Corporation, RR-2165-A, 2018. As of August 1, 2019:
https://www.rand.org/pubs/research_reports/RR2165.html

O'Mahony, Angela, Thomas S. Szayna, Christopher G. Pernin, Laurinda L. Rohn, Derek Eaton, Elizabeth Bodine-Baron, Joshua Mendelsohn, Osonde A. Osoba, Sherry Oehler, Katharina Ley Best, and Leila Bighash, *The Global Landpower Network: Recommendations for Strengthening Army Engagement*, Santa Monica, Calif.: RAND Corporation, RR-1813-A, 2017. As of August 1, 2019:
https://www.rand.org/pubs/research_reports/RR1813.html

Oberdorfer, Don, and Robert Carlin, *The Two Koreas: A Contemporary History*, New York, N.Y.: Basic Books, 2014.

Pernin, Christopher G., Katharina Ley Best, Matthew E. Boyer, Jeremy M. Eckhause, John Gordon IV, Dan Madden, Katherine Pfrommer, Anthony D. Rosello, Michael Schwille, Michael Shurkin, and Jonathan P. Wong, *Enabling the Global Response Force: Access Strategies for the 82nd Airborne Division*, Santa Monica, Calif. RAND Corporation, RR-1161-A, 2016. As of August 1, 2019:
https://www.rand.org/pubs/research_reports/RR1161.html

Pernin, Christopher G., Jakub P. Hlavka, Matthew E. Boyer, John Gordon IV, Michael Lerario, Jan Osburg, Michael Shurkin, and Daniel C. Gibson, *Targeted Interoperability: A New Imperative for Multinational Operations*, Santa Monica, Calif.: RAND Corporation, RR-2075-A, 2019. As of August 1, 2019:
https://www.rand.org/pubs/research_reports/RR2075.html

Pernin, Christopher G., Angela O'Mahony, Thomas S. Szayna, Derek Eaton, Katharina Ley Best, Elizabeth Bodine-Baron, Joshua Mendelsohn, and Osonde A. Osoba, "What Is the Global Landpower Network and What Value Might It Provide?" *Defense and Security Analysis*, Vol. 33, No. 3, 2017, pp. 209–222.

Powell, Alexander, Larry Lewis, Catherine Norman, and Jerry Meyerle, *Summary Report: U.S.-UK Integration in Helmand*, Washington, D.C.: CNA, February 2016.

Salhani, Claude, "U.S. Wants Gulf to Buy into Missile Defense System," *Middle East Times*, December 15, 2008. As of November 12, 2009:
http://www.metimes.com/International/2008/12/15/us_wants_gulf_to_buy_into_missile_defense_system

Schmitt, Olivier, *Allies That Count: Junior Partners in Coalition Warfare*, Washington, D.C.: Georgetown University Press, 2018.

Serena, Chad C., Isaac R. Porche III, Joel B. Predd, Jan Osburg, and Brad Lossing, *Lessons Learned from the Afghan Mission Network: Developing a Coalition Contingency Network*, Santa Monica, Calif.: RAND Corporation, RR 302-A, 2014. As of August 1, 2019:
https://www.rand.org/pubs/research_reports/RR302.html

Shlapak, David A., and Michael Johnson, *Reinforcing Deterrence on NATO's Eastern Flank: Wargaming the Defense of the Baltics*, Santa Monica, Calif.: RAND Corporation, RR-1253-A, 2016. As of August 1, 2019:
https://www.rand.org/pubs/research_reports/RR1253.html

Tripp, Robert S., Kristin F. Lynch, John G. Drew, and Edward W. Chan, *Supporting Air and Space Expeditionary Forces: Lessons from Operation Enduring Freedom*, Santa Monica, Calif.: RAND Corporation, MR-1819-AF, 2004. As of August 1, 2019:
https://www.rand.org/pubs/monograph_reports/MR1819.html

U.S. Africa Command, *United States Africa Command 2015 Posture Statement*, March 17, 2015. As of August 1, 2019:
https://www.africom.mil/media-room/document/25285/usafricom-posture-statement-2015

U.S. Army Africa personnel, interview with the authors, Vicenza, Italy, March 2018.

U.S. Army Europe personnel, interview with the authors, Wiesbaden, Germany, April 2018

U.S. Army Pacific personnel, interview with the authors, Honolulu, Hawaii, May 2018.

U.S. Army South personnel, interview with the authors, San Antonio, Tex., March 2018.

U.S. Department of Defense, *Summary of the 2018 National Defense Strategy of the United States of America: Sharpening the American Military's Competitive Edge*, Washington, D.C., January 1, 2018. As of October 29, 2018:
https://www.defense.gov/Portals/1/Documents/pubs/2018-National-Defense-Strategy-Summary.pdf

Vergun, David, "Army, Allies Strive for Greater Interoperability in Europe," *Army News Service*, October 18, 2017. As of July 9, 2018:
https://www.army.mil/article/195310/
army_allies_strive_for_greater_interoperability_in_europe

Votel, Joseph L., "The Posture of U.S. Central Command—Terrorism and Iran: Defense Challenges in the Middle East," statement before the House Armed Services Committee, Washington, D.C., February 27, 2018. As of August 1, 2019:
http://www.centcom.mil/Portals/6/Documents/HASCVotel20180227.pdf

Wasserby, Daniel, "NATO 'Spearhead' Force to Take Shape by February 2015," *International Defence Review*, October 7, 2014.

Weitsman, Patricia A., "Wartime Alliances Versus Coalition Warfare: How Institutional Structure Matters in the Multilateral Prosecution of Wars," *Strategic Studies Quarterly*, Vol. 4, No. 2, 2010, pp. 113–138.

———, *Waging War: Alliances, Coalitions, and Institutions of Interstate Violence*, Stanford, Calif.: Stanford University Press, 2014.

Welch, Jason, "Intel Workshop Combines Coalition and Iraqi Experiences," Defense Visual Information Distribution Service, August 13, 2018. As of October 29, 2018:
https://www.dvidshub.net/news/289581/
intel-workshop-combines-coalition-and-iraqi-experiences

Wormuth, Christine, "How Is DOD Responding to Emerging Security Challenges in Europe?" testimony before the House Armed Services Committee, Washington, D.C., February 25, 2015.